CHAMELEO

CHAMELEO

A STRANGE BUT TRUE STORY OF INVISIBLE SPIES, HEROIN ADDICTION, AND HOMELAND SECURITY

ROBERT GUFFEY

OR Books

New York · London

Published by OR Books, New York and London
Visit our website at www.orbooks.com

All rights information: rights@orbooks.com

First printing 2015

Cataloging-in-Publication data is available from the Library of Congress.
A catalog record for this book is available from the British Library.

ISBN 978-1-939293-69-5 paperback
ISBN 978-1-939293-70-1 e-book

Text design by Bathcat Ltd. Typeset by CBIGS Group, Chennai, India.

To the inimitable Mr. Watts...

...and to Richard...

...without both of whom this book could never have been written.

"I warn you that what you're starting to read is full of loose ends and unanswered questions. It will not be neatly tied up at the end, everything resolved and satisfactorily explained. Not by me it won't, anyway. Because I can't say I really know exactly what happened, or why, or just how it began, how it ended, or if it has ended; and I've been right in the thick of it. Now if you don't like that kind of story, I'm sorry, and you'd better not read it. All I can do is tell what I know."

—JACK FINNEY, *The Body Snatchers*, 1955

* * *

"…Listen, therefore, to the deposition that I have to make. It is indeed a tale so strange that I should fear you would not credit it were there not something in truth which, however wonderful, forces conviction. The story is too connected to be mistaken for a dream, and I have no motive for falsehood."

—MARY SHELLEY, *Frankenstein*, 1818

Richard Schowengerdt, founder of Project Chameleo, in March of 2006. Photograph by Melissa Guffey.

Robert Guffey interviewing Richard Schowengerdt in March of 2006.
Photograph by Melissa Guffey.

Dion Fuller in Seattle, summer of 2004, standing beside the van he drove from San Diego all the way to Winona, Kansas. Photograph by Robert Guffey.

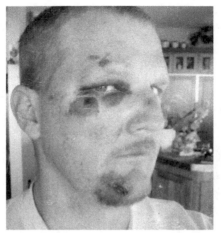

A more recent photograph of Dion Fuller.

Dion Fuller's farewell message to the NCIS and their invisible devil spawn.

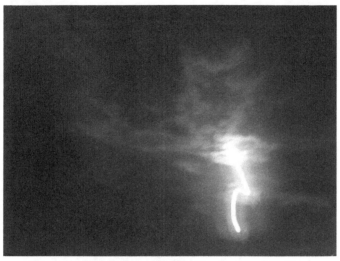

A UFO-like drone streaking off into the skies above Humboldt County in July of 2012. Photograph by Dion Fuller.

1.

My friend Dion first told me about the invisible midgets in the summer of 2003. Not long before, he'd caught his girlfriend Jessica having sex with some other dude in Pacific Beach, a suburb of San Diego, where they lived, and he physically threw her out of his apartment. Without Jessica's stabilizing influence, he sank deeper and deeper into his old habits. Pacific Beach was the worst possible place for Dion at that moment. The only people there were alcoholics and meth addicts and the policemen who arrest them.

He started drinking more and more. Then he went back to speed and meth and heroin and everything in between. One day he was riding along on his bicycle, drunk, and slammed into a building or a police officer or something like that, and broke his leg. As a result he couldn't go to work anymore at Bub's Dive Bar & Grill. Or so he claimed. It could be they just fired him for becoming more and more of a total fuck-up, something Dion had a talent for. Anyway, somehow he was able to wrangle a steady stream of disability checks out of the government while waiting for his leg to heal. With nowhere to go, he decided to start selling drugs out of his apartment, which was located at 1621 Hornblend Street between Jewell and Ingraham. That is when everything definitely took a turn into *The Twilight Zone*.

More and more fuck-ups and scumbags were hanging around the apartment. The place became notorious. The cops drove by there all the time, just to make sure nothing was getting *too* out of hand. One night, in the midst of another twenty-four hour party, some kid in his early twenties named Lee dropped by the place and asked if he could stay there for awhile. Dion's reaction was sure, what the hell, why not. The place was a party house. People were coming in and out of there all the time. What was one more person?

This kid, however, was different from all the drifters who had stayed at the apartment before. Lee had recently gone AWOL from nearby Camp Pendleton. He had taken with him 1) twenty-five pairs of night vision goggles, 2) a nine millimeter pistol taken off the body of a dead Iraqi general, 3) a DOD laptop, and 4) an entire truck. How such a feat was possible in our post-9/11 lockdown society is beyond me. The truck was not stashed at Dion's apartment. The other three items, however, were.

Lee had the goggles—three or four of the pairs, at least—stored in a trunk. Dion—perpetually buzzed out of his mind—didn't think there was anything odd about any of this until he saw the DOD logo appear on the kid's laptop one evening (on July 18, 2003, to be exact). That's when the seriousness of the situation dawned on him.

"Hey, you can't turn that on in here," Dion said. "They can track that shit with satellites! They'll be here within seconds."

Lee just waved him away. "That's bullshit. They can't do that." Dion and a bunch of other people watched as the kid scrolled through a whole series of files marked TOP SECRET and ABOVE TOP SECRET. The file names were so technical-sounding Dion had no idea what they meant. As a whole, the files seemed to be a field journal written by a team of intelligence specialists stationed in the Gulf.

Lee opened some of these files and laughed while pointing at TOP SECRET blueprints for machines Dion didn't recognize.

Finally Dion said, "Fuck this, that's it! You've got to pick up your lowjack shit and get the fuck out of here!"

Lee refused to go.

At that point there was a knock at the door. A very officious-sounding knock. All the people on drugs at the permanent party froze while Dion opened the door. A middle-aged woman flashed a badge and identified herself as Special Agent Lita A. Johnston of the Naval Criminal Investigative Services, the NCIS.[1] Two "men-in-black"-types stood behind her.

"We have reason to believe you have stolen military equipment stored on the premises," said Lita. "We're going to search your apartment."

"Do you have a search warrant?" Dion's father had been a narcotics cop for the Gardena Police Department, so Dion always knew exactly what his rights were.

After some amount of stalling, Lita had to admit she didn't have a warrant. "But we can get one in about ten minutes," she said.

"Come back in ten minutes then," Dion said and slammed the door in her face.

Dion spun around and said to the assembled partygoers, "The Feds are coming. Pick up your shit and go."

Despite the fact that Dion was always trying to convince me that drug addicts have better morals than straights, everyone at the party tossed their drugs on Dion's floor and scampered out the back like mutant rabbits. Even Lee tried to bail, but Dion pushed him back down onto the floor of his bedroom and told him if he was going to jail then the kid was going to tag along.

---

1.   If you wish to confirm the involvement of the NCIS in this most curious affair, feel free to contact me at cryptoscatology@gmail.com for further corroborating data.

Since the warrant was clearly imminent, the second everyone (except for Lee) was gone from the premises, Dion allowed the NCIS inside. Five minutes later chaos ensued as the cops arrived on the scene. The NCIS didn't know what to do. They seemed to be arguing with the cops, as if wires had gotten crossed somewhere. The cops were invading the territory of the NCIS, but the NCIS couldn't do shit about it. NCIS agents and cops kept bumping into each other like some slapstick skit from the Three Stooges. Everyone was working at cross purposes. While the cops were looking for drugs (and finding them stashed in every nook and cranny), the NCIS people were only interested in the equipment, particularly the night vision goggles. The NCIS goons began interrogating Dion and Lee right then and there, giving them the third degree, shouting the word "goggles" over and over and over again.

"Where are the rest of the goggles, where are the rest of the goggles, where are the rest of the goggles?"

Dion couldn't say because he didn't know. So they grabbed him and Lee, arrested them under the ostensible auspices of the Patriot Act, and dragged them to jail in downtown San Diego, where they gave them the Abu Ghraib treatment for six days straight. The NCIS agents had come to the conclusion that Dion was running some big-time smuggling operation out of his apartment. They accused him of stealing sensitive equipment from the ubiquitous military bases in San Diego and turning around and smuggling them over the border and selling them to foreign terrorists. The idea of Dion working up enough energy to do anything more than a slip a needle in his vein every day was sort of laughable, and I'm sure he explained that to them, but they would have none of it. If Dion had that much energy, he would've finished that damn book fifteen times by now, the one he'd been promising to write. Nonetheless, they accused him of treason again and again. They showed him photos of various people he'd never seen before and asked him, "Do you know this person?" Most of them he'd never seen before. Three of them he did recognize: the kid, Lee; a local drug dealer named Sid; and a man named Mark Hampton, also known as "Mark the Shark." Mark was a meth addict who had introduced Dion to Lee three days before the party. The NCIS told Dion that he had (and this is an exact quote) "walked into a five-year-long investigation of a smuggling ring operating out of Camp Pendleton." Lee was one of the culprits, though low-level, and the NCIS wanted to catch much bigger fish than Lee. "Lee," by the way, wasn't even his real name. The NCIS told Dion that Lee's real name was Doyle, but they didn't provide a last name—or if they did, Dion couldn't recall it when I eventually spoke to him about the situation.

The NCIS demanded that Dion rat out Lee and Mark. Though Dion didn't know much about the situation, he knew enough to serve as a witness for the NCIS in court. But he refused to do so. Ingrained in him was that honorable jailbird motto of "Never Snitch." It was something Dion took very seriously. He would go to jail himself before snitching on someone else, even if he knew the person was guilty.

After six days of non-stop questioning, the NCIS let Dion go. The police were holding a multitude of drug charges over his head, and yet (mysteriously) all such charges were dropped despite the fact that Dion could've gone to prison for many, many years based on how many drugs they found in his apartment.

Dion assumed the NCIS had come to their senses and realized he had nothing to do with the whole affair. Yes, the nightmare was over. Now he could go back to his half-life—his limbo-like existence without Jessica.

But apparently Lita Johnston and the NCIS hadn't realized anything.

This is when Mr. Big came into the picture.

Mr. Big: that's me.

2.

The same week Dion was getting the third degree in prison, I was getting a different kind of third degree in Torrance, a coastal suburb of Los Angeles.

On July 12, 2003, I became initiated into the third degree of Freemasonry at Torrance University Lodge #394. How I drifted into Freemasonry is a weird and complicated tale, best left for another time. The only reason I mention it now is because it plays an important role later on.

I returned home from the third degree early on a Saturday afternoon, called my girlfriend at the time, and told her I was still alive. She was happy to hear it, as she thought Freemasons were devil worshippers who made a regular habit of sacrificing babies to Baphomet.

The second person I called that day was Dion. The phone rang and rang. He didn't answer. Usually there was *someone* at his place who would pick up the phone, often someone I didn't know, but not this time. I didn't think anything of it. Maybe they were all passed out?

In between arguments with my girlfriend, the Insanely Jealous Poet who demanded we get married after we'd had sex for the first time only a couple of days before, I tried to call Dion over and over again. No luck. No luck. No luck. I started to get a little worried. I knew he had been upset about losing Jessica, and I knew he was doing drugs again. But, on the other hand, he had gone missing for several days in the past, so this was nothing new. I pushed the situation out of my mind and went back to my endless arguments with the Insanely Jealous Poet.

About twelve days later I received a phone call from Dion in the middle of the afternoon. He was out of breath, excited about something, and proceeded to tell me the entire story of how he had ended up in jail for the past week. My initial reaction was one of stunned amusement. The idea of Dion being the head of a smuggling ring was ridiculous. Even more ridiculous was the irony of the NCIS and the San Diego Police Department wasting their considerable resources for over a week interrogating Dion about a smuggling ring while a whole slew of illegal aliens who might have plutonium bombs strapped to their backs were pouring over the border only a few miles away, unhindered. (But stopping *that* problem would require some amount of effort, whereas interrogating Dion was something that could be easily accomplished between dough-nut breaks. Featherbedding: a true American tradition!)

Overall, Dion seemed relieved that the NCIS and the cops had finally come to their senses and allowed him to bail. He returned to his apartment to discover that it had been ran-sacked during his incarceration. Every drug addict who had ever stayed there was well aware of the fact that Dion was behind bars, so someone took the opportunity to rout through the mess left behind by the cops.

Based on his story, I didn't understand why the cops didn't charge Dion with drug possession. This seemed to puzzle me more than it puzzled Dion. Why would they just let him go?

Whatever drugs had been accidentally left behind by the cops—if, indeed, there even were any—had been swiped by the scavengers who swooped down on the place like vultures in Dion's absence. All throughout his six days in jail, while the NCIS grilled him over and over again about the current whereabouts of their precious night vision goggles, Dion kept pleading with them to put a guard on his apartment. The authorities had only retrieved a few of the goggles, so Dion thought Lee had stashed them somewhere in his pad without his knowledge. Dion understood very well what would happen to his place while he was gone. But neither the cops nor the NCIS listened to him. They kept accusing him of lying and just wanted him to confess to stealing the goggles.

It didn't make sense to me that the authorities would let him go. What had been the point of it all? Dion thought that maybe they had gotten Lee (or Doyle, or whoever the hell he was) to confess to the crime. Perhaps he broke under the pressure and told them where they could find the rest of their shit? Dion didn't seem to care about the details. He just wanted to move on.

Though all of this was odd and intriguing, my main concern at the time was that my girlfriend had tricked me into having sex with her without a condom because she claimed she was on the pill when she wasn't, so I was sweating hollow-point bullets while waiting to discover if her ploy had worked. I quickly forgot about Dion's weird drama.

Until a few days later when Dion called to tell me he was being followed. At first I thought he was suffering from some kind of meth-induced paranoia. Meth addicts are always seeing weird shit and suspecting people of spying on them, etc., so this was nothing new to me. What was different, this time, was how grand the scenario became. And how frightened Dion seemed. I'd rarely ever seen him frightened of anything.

He told me he had just visited the 7-11 on Garnet Avenue not far from his apartment in Pacific Beach. The second he entered, about a dozen jarhead-looking guys followed him in. Real military-looking types. They just tailed him around the store, right on his ass, being really blatant about it. Then he left the store and the jarheads followed him out. I started to laugh as he was telling me this, because I was imagining a whole parade of people trailing Dion down the sidewalk. But then Dion explained it wasn't like a parade at all. They would spread out and follow him in such a way that the casual observer wouldn't notice anything unusual. Dion couldn't help noticing these guys following him all around town. When he made the mistake of confronting them and asking what the fuck they were doing, they did not respond in any way. They acted like he was invisible.

What started out as a dozen people following him grew to at least *forty-five* people or more harassing him in broad daylight, right in the middle of the street. He claimed people would plant themselves outside his apartment and shine halogen lights through his windows for hours at a time.

Wow, I thought, Dion's gone right off the fuckin' deep end. It was inevitable, I concluded. It had to happen sooner or later.

But then things began to happen that made me think, well, maybe Dion wasn't so paranoid after all. Keep in mind this was 2003, the height of the Cheney/Bush administration right before their "re-election" in 2004. During the Nixon administration the homes of hundreds of anti-war activists were burglarized by The Plumbers, the President's own personal covert team of domestic terrorists led by G. Gordon Liddy and E. Howard Hunt, who did everything in their the power to get the President re-elected in 1972. And these tactics succeeded. In one case, Liddy even attempted to murder a famous investigative journalist (Jack Anderson) who had published numerous articles criticizing the President's policies in Southeast Asia. If there were any successful murders, of course, we don't know about them. Some people even believe the infamous 1971 break-in of Philip K. Dick's San Rafael apartment was perpetuated by The Plumbers.[2] This could very well be, as Dick had signed a petition refusing to pay 10 percent of his income tax in order to protest the Vietnam War. Others signers of that same petition received even less pleasant attention from The Plumbers and their ilk.

Was history repeating itself? Orwellian surveillance tactics exploded in the wake of the 9/11 terrorist attacks. The order of the day was to round up martyrs, no holds barred. Was it so impossible that an innocent (well, "innocent" at least of the

---

2. Philip K. Dick was the author of *Eye in the Sky, Do Androids Dream of Electric Sheep?, A Scanner Darkly,* and many other classic paranoid novels.

crime for which he was being accused) could get caught up in the net through a simple, stupid error? At the beginning of Terry Gilliam's science fiction film *Brazil* (Universal, 1985), an average Joe is assassinated by a government taskforce because of a simple bureaucratic error. The word "Tuttle" is transformed to "Buttle" due to the corpse of a fly mucking up the inner workings of a typewriter, and Buttle is killed for acts of terrorism while Tuttle goes free. Had Dion landed the role of Buttle in this absurd dark comedy?

Dion told me how one tall, jarhead-lookin' guy followed him around town for hours. In order to lose him Dion ducked into an Army Surplus store, then walked right back out again. The guy followed him in, then followed him back out. So Dion strolled across the street and bought a 32-ounce bottle of beer. His shadow followed him in. Dion left, turned around, and tossed the entire beer in the guy's face. The guy froze for a moment, but did nothing. Didn't say, "Hey, you motherfucker, what'd you do that for?" Just stared at Dion and let him walk away. As if he were under orders not to engage with his quarry, no matter what the quarry did. It was unnerving to Dion, so he immediately hightailed it home to call me.

In another instance, not long afterward, Dion strode right up to one of these alpha-male-looking primates, got right in the guy's face, and screamed, "Faggot!" Usually this would illicit a response—something, anything.

Nope. Nothing. And Dion did this several times.

Strange stories like this kept piling up, one on top of the other. The second he'd leave the apartment, for example, five of these guys would be on his ass—getting in his way, jostling him "accidentally," and impeding his progress in whatever way they could. Just walking to the supermarket down the street was a Herculean task.

He insisted the same exact cars were following him all around Pacific Beach and beyond. So I told him to write down as many of the license plates as possible. He'd just walk right up to the back of the car, while the driver was still sitting inside it, and meticulously jot down the numbers. After awhile he'd collected nine of these:

4U49864 (a white truck)
5DA726 (a white Sedan)
4WEP4HEP852 (a brownish-copper Honda)
SE4 JPL (a white Toyota)
5BEA709 (Escape)
4EPW827 (black Acura)
809-E2G (Toyota Forerunner)
4AMP126 (a white Toyota)
4JPL825 (Xterra SE)

In order to put all of this to the test, I gave the numbers to a friend in Seattle, who turned around and gave them to a guy he knows who works for the DMV. In his spare time, the guy at the DMV ran all the numbers. None of the license plates had ever been issued.

Stop a moment and think about this: if Dion were paranoid, and merely *thought* people were following him around, then these license plate numbers would have been assigned to real people with real automobiles. If Dion were lying for some reason, and just making up the numbers off the top of his head, the law of averages would dictate that he would conjure up at least *one* real license plate number out of those nine we checked. But not one of them officially existed. Which is what you would expect from a government vehicle, as automobiles used in covert affairs like this are never listed in the DMV files. There's no way to explain the non-existence of the license plates except the conclusion that Dion was being followed by government vehicles.

So now I believed him... kind of. He *wasn't* merely suffering from meth-induced insanity. That much had been proven. Now what? What the fuck could *I* do about it?

The only thing I could do: I took extensive notes. The U.S. government was studying Dion from one end and I was studying him from the other.

At least he wasn't lonely anymore.

3.

While attending the Clarion Writers Workshop back in 1996, I learned a phrase that describes an error many young writers make while beginning to write stories. The phrase was coined by novelist James P. Blaylock. It's known as "the Octopus On the Shelf." The Octopus On the Shelf is an element in a science fiction or fantasy story (though I suppose it could just as easily occur in a mainstream story) that causes the reader to think, "What the fuck? Where did *that* come from?" But not in a good way. Not in a way that keeps you interested in the story, but in a way that makes you question the sanity and skill of the writer.

The first Octopus On the Shelf (oh yes, there were several) occurred when Dion called me one day in a panic and told me that invisible midgets were infiltrating his apartment.

"How do you know they're midgets?" I said.

"Never mind that," he said, then proceeded to regale me with stories about people brushing up against him in his living room when no one was around, pushing him over, moving furniture around, and making a general nuisance of themselves.

I had now reverted to my old opinion that Dion had gone nuts. After all, it was more than possible that government agents were indeed watching him (because they thought he was still hoarding their precious night vision goggles) and that Dion was *also* suffering from meth paranoia at the same exact time. These were by no means mutually exclusive situations.

But as Dion kept babbling I became intrigued by the weird little details that tumbled from his mind. For example, at one point he said he was in his bathroom getting something out of the medicine cabinet when he caught sight of one of the little fuckers. In the split second that it took him to open the cabinet door, he spotted a little man in the background standing only about ten feet behind him. He didn't get a detailed look, but he saw enough to know someone was there. Dion started opening and closing the mirror like mad, but by that time the homunculus had skedaddled. I thought this was a very intriguing detail. As the science fiction writer Theodore Sturgeon once said, "Always ask the next question." If someone had an experimental invisibility suit based on light-bending technology, it made some sort of cockeyed sense to me that the suit might become visible temporarily while the mirror was in motion. Dion was hardly a science fiction fan, nor did he own a subscription to *Popular Science*. It's not the kind of detail he would make up off the top of his head. He was just reporting what he was seeing.

Another detail he noticed was that, occasionally, parts of the little people would become visible. Depending on the background, a vague outline would appear represented by dots of light floating in the air. Dion said they looked something like the auras he would see when suffering from a bad migraine. In fact, at first he thought that's what they were. But then he began seeing them more and more frequently, often when he wasn't suffering from a migraine, and only when he could sense the presence of other people in his apartment. I had no idea what this detail indicated. The flashes of light made me think of visual phenomena that some people report while hallucinating. Again, there was something about this detail that seemed authentic to me. I couldn't explain why, however, so I just put it in my "gray basket" (neither positive or negative, just gray) for the time being and promised myself I'd try to follow up on it later.

Meanwhile, Dion kept insisting on referring to his visitors as "invisible midgets." This was partly an example of Dion's dark humor, but also semi-serious. He seemed to think they were very diminutive people. If you were going to use these super-suits in order to infiltrate sensitive military locations in other countries, you *would* want to train small and agile people to use them. Gymnasts, perhaps? Dancers? Jockeys from the nearby Del Mar Race Track? Dwarves? Why not? ("Naval Intelligence In Need of Vertically-challenged Individuals for Sensitive Intelligence Work. Must Be Willing To Terrorize Harmless Drug Addicts 24/7. BYOB.")[3]

Not long after the invisible midgets started showing up at the apartment, Lita Johnston made a return appearance. Early one afternoon, all the terrorist madness just ceased. The jarheads who were parked outside drove away. "Huh, maybe they gave up," Dion thought. About twenty minutes later, however, Lita herself appeared on his back doorstep with two male agents in tow. She was polite and friendly. She asked Dion how he was doing. He said he was okay. She then asked him if he'd changed his mind and would like to tell her where the goggles were hidden. He insisted he didn't know. She said

---

3. Improbable? Perhaps. But consider the following article by Justin Davenport, crime reporter, originally published in the Nov. 10, 2011 edition of the *London Evening Standard*:

    "A London police force is being sued over claims that it used two undercover dwarfs to carry out an anti-terror search.

    "A Russian doctor is claiming 55,000 pounds saying City of London Police officers also sexually assaulted him and took his DNA to carry out 'covert biological experiments.'"

    "Dr. Alexander Sobko, of South Kensington, claims the 'smiling' dwarfs approached him at a bus stop and searched him under the Terrorism Act. The case will be considered by a High Court judge and the City force has engaged lawyers at taxpayers' expense to defend the action.

    "A City police spokesman said: 'We have instructed lawyers to contest the allegations.'"

that was too bad, because if he did know she could probably help him out.

At that point Dion just came straight out and asked her if she had people tailing him. She laughed and said that was silly. She then gave him her business card and told him if he changed his mind he could call the number at any time of the day or night. He said he couldn't call the number because he didn't know anything. She just nodded, and all three of them left in their unmarked, government vehicle.

A few minutes later all the chaos re-ensued. The flotilla of spies resumed their positions, invisible people started moving furniture around the apartment, and jarheads camped out in the supermarket parking lot next door and blatantly took photographs of him. All day, all night. Some of these guys would remain parked in the lot outside the apartment and play loud music for hours while beaming their headlights right through his bedroom window. Strangely, none of the neighbors seemed to mind. It's important to note that since his arrest in July almost *all* of Dion's neighbors, one by one, had moved out of the building. All the apartments were now occupied by new people, people Dion had never seen before.

The landlord started acting more and more nervous and kept asking Dion—seemingly for no reason—if he was planning on moving soon.

One time Dion asked him, "Is there any reason why I *should* move?"

Perspiring like mad, the landlord said nothing and crept back into his apartment.

Then the room started growing. Dion called me one afternoon—several months after this whole mess had begun—to tell me that earlier in the day he had entered his apartment through the front door and was surprised to see that the living room appeared far larger than normal. "This house is evil," Dion told me, "and it *grows!*" At this point he (and I) thought he had definitely lost his shit, until a couple of days later when his friend My Lai came over and said, "Say... does the place look, like, *way* bigger to you?"

Again, I didn't have the technical knowledge to explain such phenomena, but I knew it was not a typical Dion-hallucination. I had known him since we were both sixteen, and by this point I was used to how he behaved while under the influence of a panoply of different drugs. None of these symptoms were familiar. I've known a lot of drug addicts in my life, and not one of them ever hallucinated that their apartment had transformed into a tesseract house; no one ever shoots heroin and dreams that their domicile has become the stand-in for Doctor Who's Tardis. Hell, heroin and meth aren't even hallucinogens. Dion despised hallucinogens and always had. His personal kinks were heroin and speed, and that was pretty much it. All those mellow hippie drugs just put him in even worse moods than normal.

Again, I wasn't sure how one could make a room look bigger on the inside than it appeared on the outside, but my intuition told me it wasn't impossible.

Then came the Boris Vallejo virtual reality mindfucks. One day Dion glanced out his living room window and noticed that the scenery that *should've* been there had disappeared. Instead he saw what he described as the background of a cheesy Boris Vallejo painting: swirling green mists, alien vegetation, three suns in the sky, everything but the furry thoat and the big-breasted woman in the loincloth. I thought it was significant that the scene didn't even look like a Frank Frazetta painting; it looked like a *Vallejo* painting. Leave it to a U.S. intelligence agent to pick the second-rate cheesy painter instead of the first-rate cheesy painter for their little virtual reality scenario (it's important to note that this scene didn't appear to be a mere two-dimensional painting—it looked *real*, as if Dion could open the door and step into it).

After seeing this Vallejo Wonderland, Dion ran over to the windows on the other side of the apartment and saw the same things he'd always seen: the grocery store and the gas station and the power line next door. No weird vegetation, no green sun, no swirling mist.

He went back to the other side and glanced out the window again. Nothing had changed: Vallejo everywhere. When he opened the door, however, he saw exactly what he was supposed to see. Only when looking out the windows on one side of the building did the scene look surreal. The scenery remained this way for a few hours, then morphed into something else. This continued for several weeks, driving Dion nearer and nearer to The Edge.

Late one night, fearing for his life, he wrote out a huge sign that read "PLEASE DON'T SHOOT ME" and hung it in his front window. He crawled into bed and within seconds saw a shadow projected onto his bedroom wall on the other side of the room. It was the silhouette of a giant hand gripping a gun, and the gun was pointed at the silhouette of his own head. It kept tilting up and down, aiming at his forehead and pulling the trigger. Over and over again.

* * *

The strange occurrences persisted as the weeks turned into months, from July of 2003 to January of 2004. About seven months. Now, you might be asking yourself: Why didn't Dion just leave? This question often comes up in regards to haunted house scenarios. How come these idiots don't just bail if it's so damn scary? Well, when you don't have a lot of money it's kind of hard to leave. Also, Lita continued to warn him over and over again *not* to leave the city.

Yes, Lita continued to show up at his door from time to time with the same two agents in tow, asking if he'd found the damn goggles. He kept saying no, of course, since he'd never even seen them in the first place.

But it was around October that Dion began doing his own detective work in Pacific Beach. He went to all the thugs he kept for friends and asked if any of them knew what Lee/ Doyle had been up to. Had he really been running some kind of smuggling ring out of Camp Pendleton? Dion uncovered some scuttlebutt that indicated Lee/Doyle had stolen the goggles in order to sell them to the Hells Angels. At first this might sound ridiculous, but the Hells Angels are involved with a great deal of drug smuggling in San Diego and Mexico. A supply of hi-tech night vision goggles is exactly something you'd need to make your job easier while smuggling major loads of contraband over the border at night. However, it's not clear to me that Lee/Doyle understood the severity of his crime. It's not possible that those night vision goggles were just normal everyday goggles. They couldn't be. You could repurchase twenty-five pairs of those babies through eBay for far less money than it would cost to keep this psychological warfare game going against Dion Fuller.

4.

Dion never really needed a good reason to try and kill himself, but now he was being given one on a silver platter. His latest suicide attempt began with a food fight.

One afternoon two jarheads were following him around the city. A nearby Mexican restaurant that had been failing for years experienced a sudden upsurge in business thanks to all the military types who would eat there whenever Dion decided to drop in and purchase an enchilada combo dinner. All of these agents were apparently on rotation. Dion never saw the same shadow twice. Two annoying shadows had been up his ass throughout the entire weekend. They were particularly brazen, and would camp out on the other side of a wooden fence outside his kitchen window. While he was standing there making breakfast or lunch or dinner, these guys would peek over the fence and try to take photos of him and observe his activities through a pair of binoculars. Their purpose, no doubt, was just to annoy him. Imagine a ridiculous college fraternity with the resources of the entire black budget of the United States of America deciding to play one long prank on some faceless guy in San Diego. And imagine that the faceless guy is *you*. It might seem absurd at first, until you start to lose your mind.

The military types who were parked in the lot outside never left. One time Dion got fed up and called the cops just to see what would happen. One of the officers, some young kid, seemed interested in Dion's story while the older cop kept implying that the gentlemen who were parked outside weren't really there at all. The young kid just looked confused, as if one cop was in on the joke and the other wasn't. Or perhaps it was just a classic good cop/bad cop routine. Who knows? The point is, Dion could seek no help from local law enforcement as they had clearly been told to allow this operation to continue.

So one sunny afternoon, while Dion was making lunch, he glanced out the curtains and saw (sure enough) two of his most recent shadows peeking at him through the slats in the wooden fence right outside the kitchen window. By this point (early November) Dion had become resigned to these terrorist tactics. Like Number 6 in Patrick McGoohan's TV series *The Prisoner*, and of course Winston Smith, Dion had grown accustomed to being under constant surveillance. He had no other choice. He would sleep, eat, fuck, whatever, knowing someone was either watching him or recording it all for posterity. In fact, there were several times when the invisible midgets would make their presence known while he was jacking off in the bathroom.

As his two shadows continued to spy on him, Dion non-chalantly went about his business. He started making spaghetti, but as the constant sight of these two jarheads began to grate on him more and more he started adding special ingredients to the mix: shredded cheese, Worcestershire sauce, rice, flour, mustard, ketchup, cinnamon, cayenne pepper, peanut butter, Tang, Jell-O mix, sugar, honey, multicolored sprinkles, and stale almonds. He stirred it all up until it was a nice, thick, noxious goo, then ran outside, tossed the entire concoction over the fence, and hit at least one of the agents dead on. Both of them screamed and went running off toward the nearby Vons parking lot.

Satisfied, Dion brushed his hands together, went back inside, and made lunch: a bologna sandwich.

For a couple of hours the surveillance lightened up a bit, but this reprieve didn't last long. By nightfall the terrorist tactics had been amped up to eleven. Reluctantly, Dion dug Lita's phone number out of his pocket and decided to make a deal with the Devil. He would at last violate his principles and cooperate. He called her cell phone at five in the morning. She picked up almost immediately, as if she were waiting for the call.

"Yes?"

"Hi, I want to talk to you about your goggles. I can't take this anymore."

"Take what?"

"You know damn well what. The harassment. Being followed."

"Who's following you?"

"You're following me."

"That's ridiculous. Why would I be following you?"

"Okay, right, fuck it. Were you asleep?"

"Yes, I was."

"I'll talk to you later."

He hung up. She never called him back. Instead, she showed up at his doorstep with more junior partners in tow a couple of days later and said, with a smile on her face, "So... you still think you're being followed?"

By this point Dion had decided he didn't want to cooperate anymore. He said, "No. I don't think I'm being followed. I was mistaken. Confused. Sorry."

She then launched into a panoply of questions about the goggles, about how many had been taken off the base, etc.

Once again Dion said, "Listen, I don't know! I knew this guy three *days*! I have no idea!"

Lita said, "Okay, if that's the way you want it," and left. At that moment, sitting in Dion's apartment was none other than Mark Hampton (Mark the Shark), the guy who introduced Dion to Lee in the first place. Word to the unwise: it's never a good idea to approach a possible informant in front of the *very people you want that possible informant to snitch on*. Apparently Lita either didn't know this simple truth or didn't care.

Dion had been talking to Lita on his front porch, and when he turned around to walk back inside he realized he'd locked himself out of his own apartment. So he had to go around and enter through the front door. When he did this he saw Lita and her partner get into a government vehicle and drive away followed by two of the cars that had been tailing him for the past few months.

The harassment continued to escalate for the next few weeks, so much so that Dion again considered calling Lita and reaching some kind of agreement. But before he could do so, to his surprise, Lita called him herself early one evening and asked him to meet her the next day for lunch at a café on Garnet Avenue, not far from his apartment. When she said she was buying he agreed, but asked if they could meet somewhere farther away since he didn't want to advertise to everyone he knew that he was talking with the Feds. She insisted, however, that they meet at that particular café.

Dion hung up and called Mr. Big (me). "I've got a plan," he said.

"Yeah?" I said. "What is it this time?" All his previous plans had failed miserably—not just in regards to this situation, but in regards to *all* situations across the board. My hopes weren't high. "I'm having lunch with Lita. I'm gonna make a deal with her."

"A deal? What're you talking about? You don't have anything to deal *with*."

"Maybe nothing physical, but look at it this way. I know a lot of people in San Diego. I know the *right* people. I know people who would never talk to Lita Johnston even if they had bamboo shoots shoved under their fingernails. Maybe I can put to use some of those detective skills I picked up from my dad and track these goggles down to the source. I've already got a lead about these Hells Angels. Maybe if I help them get these things back they'll get off my neck and… well, who *knows*? Maybe they'll offer me a job. I could work the other side of the fence for a while. I've got the expertise. I've got the skills they need."

"That's the dumbest fucking plan I've ever heard," I said. "If you offer to work for them, if you request to be put on the payroll in order to get back their goggles, they're going to assume you had the damn things all along and now you're just using these things as leverage to get money out of the U.S. government. They're not going to interpret this as a friendly act of desperation. This is just going to confirm what they've suspected all along: that you've been *holding out on them* this entire time. They're going to interpret this as a shakedown. What skills could you possibly possess that the U.S. government doesn't already have access to?"

Dion just scoffed at me. "They don't know what they're doin'! Obviously! If they knew what they were doin', they would know I don't *have* the damn things."

"I wonder if that's even the issue anymore."

"What do you mean?"

"Listen," I said, "my first teaching gig was as an English tutor at El Camino Community College. You know how I got the job? I walk into my counselor's office one day, to look over my grades, and she suddenly says, 'Hey, your grades in English are pretty high. Would you like to become an English tutor here?' I go, 'Uh… sure.' She says, 'You see, every year the government gives us a certain amount of money and if we don't spend it then they'll take it away from us next year. So we've got to figure out a way to spend it. Now, we don't really need an English tutor. I mean, we don't have much of a demand for one. So pretty much no one's going to come in. You can just sit in the back room and do your homework or read or whatever and you'll collect a pay check. I mean, we've got to unload this money somehow, so why not on you, right?' I go, 'Uh… sure.' And she sticks out her hand and says, 'See ya next week!' That's how I started teaching people English."

There was a moment of silence on the other line, then Dion said, "What the *fuck* does that have to do with anything?"

"It has to do with the reason they're still following you. Maybe they *know* you don't have the goggles. Maybe they're spying on you because they have a certain amount of money they have to spend every year in order to help stave off international terrorism, but unfortunately there's no one to fuckin' spy on so they have to pick someone who kind of fits the profile, at least a little bit, enough to justify the expense, but not enough so they have to do any real work. So they decide to pick on someone who stumbled into their hands one day and pissed them off and that happened to be your sorry, stupid ass."

"Hey."

"It's *true*. You should've just told them everything they wanted to know when they first asked you."

"Look, I'm no snitch. I told them everything they needed to know, I just didn't want to rat on my friend."

"This 'friend' of yours was the guy who got you in this mess. You knew him for seventy-eight hours!"

"Listen, you learn in jail never to snitch. It's a code."

"It wouldn't have been *snitching*. You simply would've been telling them what happened. That would've required pointing your finger at the guy responsible. What's wrong with that?"

"Fuck the pigs!"

"Right. 'Fuck the pigs, but give me an application.'"

"Hey, this is totally different. I won't be snitching. I'll be doing a *job*. They'll be hiring me for my special services. That's totally different."

"If you find out who has the damn goggles, don't you have to tell them who's responsible? How's that any different from snitching?"

"That's just fulfilling a contract. That's honorable."

"So the difference between snitching and not snitching is that in one situation you're getting paid and in the other you're *not*?"

"That's right."

"So you want them to pay you off."

"It's not a payoff. I don't even have the *information* yet. Once they pay me, then I'll go and find out what they want to know. That's a fair exchange of information, not snitching."

"I'm confused. You're still working for The Man."

"It may look that way. I only work for myself. I'm free-lance. I don't even care about the money. Dude, I just don't want to be harassed anymore! This is getting ridiculous. I'd be laughing about it if I weren't on the verge of having a fuckin' heart attack!"

I could hear the tension in his voice. It really sounded as if he were about to snap.

"Listen," I said, "just try to relax. You need your rest. You have to be alert when you talk to them tomorrow."

He agreed and hung up. A few hours later he called back, hopped up out of his mind. Apparently it was all too much for him. In order to deal with the stress he'd decided to shoot up enough heroin to kill thirteen elephants. I should never have challenged his ethical code. He couldn't handle that. Imagine telling Robin Hood he was nothing but a fucking thief… and imagine that this was *true*, but Robin Hood didn't know it until you mentioned it to him.

Dion told me he felt like he was going to die. And his voice had changed. I could hear it. Death was near. I'd been around him plenty of times when he had shot up way too much heroin, but he'd never sounded like this. For the first time, I was genuinely worried that he would die. I told him I was going to hang up the phone and call 911, but every time I suggested this he would plead with me to stay on the line. "I don't think I can stay awake unless you keep talking to me. I don't know what's going to happen if I fall asleep…" Then he'd trail off and go silent. I'd have to hit the buttons on the phone to shock him awake again. Then he'd say, "Talk to me. Talk about something. Anything." I'd start blathering about invisibility or the NCIS or international terrorism or Lita Johnston and he'd say, "No, no, anything other than all that shit! I can't take it!" So then I'd have to act like nothing unusual was going on and make up some completely mundane topic and talk about that. I'd talk about my misadventures with the Insanely Jealous Poet, the latest Monday night ritual down at the Masonic Lodge, the story I was writing at that moment, the article I had just sold, the novel I was halfway finished with, the classes I was teaching that semester, and then he'd tell me to stop because he was becoming envious and thought my life was far more interesting than his.

"How could that *be*?" I said. "You're being attacked on a daily basis by invisible midgets and having food fights with Naval Intelligence. How could that be boring?"

"But what does it all *mean*?" he moaned. "What have I accomplished? My writing's going nowhere."

"You managed to sell that one story."

"Yeah, to a *porn* magazine."

"You made more money off that one story than most people make off three."

"But you had to proofread it for me."

"Well, that's true."

"Jessica! Where are you? Why did you leave me?"

"You threw her out, remember?"

"Why did I do that?"

"'Cause she was fucking some dude behind your back."

"Why'd she *do* that to me?"

"I don't know."

He started to cry. "This is it, man. I can feel it. I'm slipping away…"

Then he went silent. I had to hit the buttons again to jerk him awake. This process went on from about 1:30 A.M. to 5:00 A.M. when he suddenly faded out again, and no amount of button pushing on my part would wake him up. The phone line went dead. I tried calling some friends of his in San Diego to ask them to go over and check on him, but none of them were home. Despite my extreme worry, I was exhausted from my marathon telephone conversation, so I just went to sleep and hoped for the best.

I slept for five hours. I woke up around noon and immediately tried to call him. The phone just rang and rang. I wasn't sure what I should do. A few minutes later the phone rang. It was Dion, sounding spry and healthy.

"Where the fuck have you *been*?" I said.

"I just got back from my meeting with Lita," he said.

"How'd you make that meeting? I thought you were dead!"

"So did I, bro, so did I. I woke up out of the nod around 11:30, got dressed, and made it to the restaurant just in time. Get this: Lita and her boss looked like hell. They looked way worse than *me*, like they'd been up all night."

I laughed. "They were up all night listening to our phone conversation. They thought you were going to die. *Then* where would they be? No goggles, no nothing."

"Exactly! So I wake up and leave the apartment and guess what I see? Stuck in my neighbor's screen door is a business card that says, 'Bob, Please call me. Special Agent Lita Johnston,' with her cell phone number scribbled on the back."

"What the fuck?"

"That's exactly what I thought. So, I got there and saw Lita sitting in a booth near the back. There was this older guy there, some fat dude. She introduced him as her superior. They bought me a bagel with cream cheese and an orange juice. And Lita was scarfing down a bean salad. So I sit down and we start off with some small talk. I ask them how long they've been living in Pacific Beach. Turns out they've lived here all their lives. And Lita says something like, 'This used to be a nice neighborhood, then a lot of criminals started moving in and taking over everything.' And I said, 'Yeah, I know what you mean. This whole town would be better off if we could figure out some way to get rid of the scumbags. I know what you're going through. I have to deal with them all the time.'"

I found it ironic that Dion was not aware Lita had been referring to *him*. He was honestly being sincere and complaining about all the scumbags in Pacific Beach he had to deal with on a daily basis while going about his law-abiding career of selling illegal substances to his friends and neighbors. I can only imagine Lita's frustration mounting as Dion went off on one of his patented tirades about being The Last Honest Man on Earth.

"Let's get down to business," Dion finally said. "Are you going to stop following me?"

Lita's boss said, "I can assure you, Mr. Fuller, no law enforcement agency is following you."

"That makes me feel a lot better. So I'm being followed by people who *aren't* employed by a law enforcement agency? Look, I don't care who they work for, I just want it to stop. How about it? How much are you willing to pay me to get these goggles back? Now, I can't promise anything, but I'm willing to try. But first you've got to get these non-law-enforcement-agency types off my back. I can't break into someone's house with forty-five jarheads on my ass. You've got to call 'em off if you want any results."

Lita and her boss then offered to put him up in a secret hotel room somewhere far away if he cooperated with them. "I don't need a secret hotel room," he said. "I'm not afraid of Mark the fuckin' Shark, I'm afraid of the jarheads who've got their crew cuts up my ass 24/7. I need the harassment to stop so I can help you. Oh, and by the way, while we're on the subject, what would the paycheck look like? I'm talkin' on a weekly basis. I don't like to be paid by the month. I know a lot of government gigs are like that, but that's not for me. My friend's a college teacher and he gets paid by the month. Fuck that. I have a life to live. I don't want to make one paycheck stretch out for four whole weeks."

Lita and her boss just stared at him for a few seconds, then Lita said, "Here's how it goes. You see, you're the one in danger of being tossed into prison for twenty years for stealing government property. You don't get to make deals with us. *We* tell *you* what you're going to do. And you're going to tell us where the fuck those goggles are."

Dion slammed his fists on the Formica tabletop. "I don't fucking know where they are! Don't you think I'd tell you by this point? Who'd go through all this just for a couple of goggles?"

"A couple?" Lita said. "Where did the other twenty-three go?"

"A *couple*. It's just an expression. A few goggles. A couple dozen. Whatever. I've never even seen them. Look, forget the money. I'll help you try to find the things if you just call off your dogs. I don't want to live like this anymore. I just want to go back to the semi-boring life I had before I ever even saw you people. I don't want to have anymore food fights with the Feds!"

Lita and her friend exchanged furtive glances and then laughed. Lita said, "Yes, that was funny. You gave us all a few giggles with that one."

Note: That comment from Lita was the only time during this whole cat-and-mouse game that Lita ever admitted, even slightly, that they did indeed have Dion under surveillance. It was the one moment in which the innocent, Columbo-like façade cracked. She really shouldn't have said it, as that single admission stirred some last lingering flames in Dion's spirit. He pulled out the card he'd found in his neighbor's door and said, "Can you fuckin' explain this shit?"

Lita grew flustered and said, "That card wasn't intended for you."

"I can see that," Dion said. "First of all, my neighbor's name isn't Bob, it's Ben. Does he even live there anymore? I haven't seen him in weeks."

Lita's superior grew enraged all of a sudden and said, "Listen, we don't believe a single word that's come out of your fuckin' mouth! We know you've got those god damn goggles. Give them to us now or we're going to arrest your ass."

"You know what? Fuck this shit! You should be *begging* me for my help. I don't need those goggles but you people apparently do. For what I have absolutely *no* fucking clue, but I can guess it has to do with some real perverse shit and I don't have the Need To Know, and guess what? I don't even *want* to know. So when you want to talk terms, just call me and maybe my Hells Angels friends will be kind enough to give you your toys back. *Suerte!*"

Then Dion got up and started to walk out, and as he did so Lita's boss turned red with fury and yelled, "I hope you enjoyed that breakfast on tax payers' expenses!"

This grand breakfast was nothing more than a plain bagel with a couple of blobs of cream cheese smeared on it and a tiny glass of orange juice. "Wow," Dion said, "I'm sure that half-eaten bagel really pushed the tax payer into the hole, not that continental breakfast you're suckin' down your gullet like a milk shake. Or *the thousands* of toy soldiers you have to pay to follow me around for no reason. Which one of us is really fucking over the tax payer, old man? Besides, I worked as a fry cook in this crazy town for three years. My tax money was payin' for *your* ass to sit around and do nothin' except harass people who are smarter than you are. I want *my* money back. When you want to give it to me, you know where you can get ahold of me, Junior. See ya!"

Then he walked out.

And he went home and filed a report with Mr. Big again. As always, Mr. Big took extensive notes.

"Wow, that was a good idea," I said. "Your plan went smoothly. I'm sure they're going to react kindly to your behavior."

"Well, the jarheads aren't outside right now. Maybe I managed to talk some sense into them."

"Or maybe they're giving you a false sense of security before they pull the rug out from under you again, like they've done a billion times before. It's a pattern. They're following some psychological warfare text book that was written back in 1947. They tend to stick to the same old pattern over and over again. You can set your watch by it. You haven't noticed that?"

"No, I guess I've been too busy trying to find the little piece of cheese in the middle of the maze."

"Do you know how many mice had to die before they figured out the perfect method of driving Dion Fuller crazy?"

Dion laughed. "It's amazing. If they put all this money and effort into solving homelessness or world hunger, we'd be living in a utopia right now."

"Utopias aren't cost efficient," I said. "Dystopias run on fear, and that's the only real product America's an expert in manufacturing these days. Maybe it wasn't always that way, but that's the way it is now. I think you need to get some sleep, lay low for the rest of the day, and just see what happens. Besides, I need to go grade some papers."

"I just want to end it all. Hell would be way easier than this shit. Where's Jessica?"

"Jessica's gone. Forget about her. You've got to deal with this shit *now*, and I've got to go grade some papers." I'd already wasted last night and half the day with this situation. I didn't need to give Lita and her cronies more content for the files they were sending over to The Central Scrutinizer.

While Dion was in jail they kept wanting to know who Mr. Big was. They kept hounding him: "Who're you working for? Who do you answer to? Who gives you the orders?" He kept saying, "Listen, I don't answer to anyone. I don't even know anyone! My social life went down the toilet the second my girlfriend left me!" "Then who the hell were all those people in your apartment when we arrested you?" "How the fuck should I know? They just live there from time to time!" "What're you, a comedian?" "It's a fucking party house! What don't you understand about that? All kinds of riffraff run in and out of there. I barely know any of them!" "You expect us to believe that someone was living in your house with equipment stolen from a military base but you weren't aware of it?" "I wasn't aware of it until about two seconds before you showed up on my doorstep!" "Then why didn't you let us in?" "You didn't have a search warrant!" They just stared at him as if he were crazy. Dion said, "So you don't party very much, do you?" From the look on Lita's face, he knew the answer was no.

So, of course, the second he got out of jail, who does he call? The only person he ever called when he was in trouble: me. He was calling me twenty-four hours a day with specific updates on the situation. And there I was asking him for exact names and dates for my notes. From the perspective of the Feds, how could they come to any other conclusion? Pretty soon I realized *I* was Mr. Big!

5.

As Dion's stories began to grow stranger and stranger, I began to grow more and more paranoid. I started obsessing on details I never would've noticed before. I began to wonder why that particular crazy person started talking to me on the bus. Were *They* checking up on me? One night I went to my favorite Mexican restaurant—La Capilla in downtown Torrance—and ordered something to go. After about fifteen minutes the waitress behind the bar called out my name: "Robert!" I picked up my enchilada dinner, paid for it, then headed for the door. The man in front of me held the door open for me—a rare, polite gesture. For a moment I was pleased. Until the man grinned and said, "Have a nice day, *Robert*!" All the tension resulting from the insane stories Dion had been filling my head with for the past three months poured out of me in an instant. Holding my enchilada dinner protectively against my chest, I pushed the son of a bitch away from me against the glass door and said, "How'd you know my name? *Tell me*!" I don't consider myself to be a particularly intimidating person, but this dude had a genuine look of fear on his face. He held his hands up in the air and said, "Hey, hey, hey, I heard the waitress say your name, that's all! It's cool, it's cool." I felt embarrassed. "Oh," I said. With my gaze locked on the ground, I turned and walked away. After I had wandered a few blocks, some drunk-off-his-ass, crimson-faced, barrel-chested, middle-aged dude emerged from The Crest Bar on Cabrillo Avenue and started talking at me. He introduced himself as "Chuck," asked if I had some spare change, to which I said no, then barraged me with all these peculiar questions: Say, didn't you go to Torrance High? Are you still in contact with anyone you went to high school with? What's your phone number? Do you ever watch porn? Do you ever masturbate to porn? Do you do drugs? Could you call my grandmother from this pay phone over here, pretend to be my old high school teacher, and plead with her to let me move back into my old room and convince her I'm really a good person at heart and that I've decided to lay off the alcohol?

I found myself actually calling his grandmother for him, partly for fun, partly to see where all this would lead. He insisted he had gone to the same high school as me, but I never remembered seeing him before. He would "reminisce" about the teachers at Torrance High, but never add any details of his own, just nod or agree with what I said. I'd make some sort of observation about Torrance High back in the late '80s, then he would repeat what I had just said, though in a somewhat different way. He did this so smoothly, I never would have noticed it if I hadn't already been on my guard because of the La Capilla incident. He begged me to let him buy me a drink at the bar he had just left, but I told him I didn't drink (he seemed shocked and confused by this revelation). When he asked me for my phone number for the hundredth time, I gave him the number of my friend Wanda, who had been Dion's girlfriend way back in high school. Chuck studied the scrap of paper on which I had jotted the number, then said, "Is this *really* your number?"

"Of course," I said, then tried to break away from him as fast as possible. He shook my hand and thanked me for all my help. I told him I hoped his grandparents reconsidered and let him move back in. He dropped to his knees and puked up all over the curb. I left him there like that as he called out for help. I thought to myself, *What kind of amateurish trap is this?* and ran away while the man lapsed into convulsions.

I called Dion right after this encounter and told him about it in detail. Dion was convinced the guy was some kind of Federal plant. Apparently the man's mission was to entrap me—get on my good side so he could pump me for information about Dion and the true location of the goggles. They assumed they knew what type of person I was simply because I knew Dion. Birds of a feather and all that. After all, Dion would've let the guy buy him a drink, no doubt about it. Six hours later, Dion would've woken up in a jail cell somewhere with false evidence planted on him. Mr. Big, on the other hand, was too quick-witted for them.

Ever seen a late 1950s film noir set in San Francisco called *The Lineup?* Throughout the movie these three thugs keep referring to Mr. Big, but none of them knows what Mr. Big looks like. At the end of the movie one of these murderers finally gets a look at Mr. Big and it turns out to be some old dude in a wheelchair. The old dude says nothing to the thug except for a couple of sentences: "You're dead. No one ever sees me." The thug gets angry and pitches the guy over a third-story railing. Mr. Big and his wheelchair land on some little boy skating on an ice rink and kill him.

I didn't want to end up like Mr. Big from *The Lineup*, so I decided to lie low. I figured I would wait until the Feds brought Dion into custody, then I'd make my move and retrieve the goggles from their hiding place after the heat had died down.

Wait. What the fuck was I *thinking?* I suddenly started to believe I *was* Mr. Big. If I went on a stroll to Fox Drug in downtown Torrance to pick up some manila envelopes in order to mail out my latest manuscript, I'd glance over my shoulder all the time. I began to notice patterns I'd overlooked before. The same person would follow me for blocks and blocks, then vanish and a new shadow would take his place. Then the old shadow would reappear and follow me back home. I wasn't sure if I was imagining this or not. I called my ex-girlfriend Stephanie, who lived in Seattle, and told her all this. She assured me I was crazy and told me to get back to work. I called my friend Eric, who lives in San Pedro, and told him what was going on. He encouraged me and Dion to buy a camera and document these spies for future reference. I called Dion and he begged me to let him come live in my apartment for awhile until the heat died down. I told him no because the heat was never going to die down. *He* was the heat! He brought heat with him. If he moved into a fairly upscale neighborhood, within weeks it would be transformed into a sewer-city filled with drugs and prostitution. All the upscale citizens would flee in terror. Sociologists called it "Dion-flight," i.e., the phenomenon of all sensible people fleeing from Dion and his nefarious spoor. Obviously, Lita was trying everything in her power to hold on to her precious neighborhood. At this point it was still up in the air as to who would win this game.

* * *

Dion was infamous for staying over for a couple of weeks that would somehow extend into a lifetime. He was the original Man Who Came to Dinner. George S. Kaufman would be baffled by Dion's parasitic powers. I've known people to flee their own home in order to get away from Dion and his leeching abilities. When Dion was eighteen he moved in with my friend Fred. Of course, he told Fred he was only going to stay for a couple of days. Fred should have known better. When Dion was sixteen he moved into my bedroom for what was supposed to be two weeks and ended up staying a whole year!

Dion quickly drove Fred right up the wall. Even though Fred was a scumbag heroin addict who would do anything for a fix (just like Dion), he couldn't stand Dion. Perhaps they were too similar. Unlike Dion, however, he was passive-aggressive. He couldn't hold his own in a fight or any kind of one-on-one confrontation. Instead of just telling Dion to get the fuck out, Fred decided to avoid the situation. The next time Dion came knocking at the door, Fred ignored him and pretended as if he weren't at home. Picture this: Fred sitting on the couch watching TV while Dion is pounding at the door and yelling for Fred at the top of his lungs. Eventually the pounding stops. A normal person would just walk away at that point. Not Dion. He decides to crawl in through the window in the kitchen. A few minutes later Dion comes strolling out of the kitchen and sees Fred just sitting on the couch watching an episode of *Beverly Hills, 90210*. Dion, having broken into Fred's apartment, is standing in the middle of Fred's living room and lecturing Fred about how fucked up Fred is and how he has no morality and should help a friend when he's down, etc. Instead of calling the cops, as you or I would have done, Fred decides to flee his own apartment! After he's gone, Dion rolls the TV out into the hallway, turns it up as loud as it can go (which is very loud), breaks off the volume knob, then locks the door behind him and leaves. The only way to turn down the TV is to unplug it, but the other end of the plug is inside the wall socket and can't be removed unless someone breaks down the door. According to Fred the TV was still like that when he returned twenty-four hours later. The Manager almost threw him out for disturbing the other neighbors.

So I can understand why Lita and her boss looked like death warmed over with butter on top when Dion came sauntering into the café. They probably thought he was dead, just like I did. I don't think they understood how Dion's mind and body could take that much abuse and still go on. Who knows what the record is for withstanding a sustained psychic attack by NCIS-sponsored invisible midgets, but I doubt it could be seven whole months, which is how long this adventure lasted.

6.

Odd coincidences cropped up that were so startling they seemed to fall far outside the realm of mere chance or happenstance. One night I punched in the terms "9 mm Iraqi gun," "night vision goggles," and "Camp Pendleton" into Google to see if any newspapers had covered the NCIS investigation of this supposed smuggling ring. I found this article by Edward Sifuentes from the *North County Times*, a local San Diego newspaper, published only two days before Dion was taken to jail by the NCIS:

LOCAL MARINE CHARGED WITH BRINGING HOME IRAQI GUN

OCEANSIDE—Military prosecutors have charged a Camp Pendleton Marine sergeant with allegedly bringing home from the war an Iraqi-made, Tariq 9 mm pistol.

Sgt. Andrew Keeler, who returned from Iraq on May 18, said Tuesday that the government is charging the wrong man. If Keeler is convicted, he could face a year in prison, reduction of rank, loss of military benefits and a bad conduct discharge, he said.

"I'm feeling pretty beaten down, I'm feeling pretty depressed... almost like this is a nightmare that I don't have any control over," Keeler said.

Camp Pendleton public affairs officials reached by phone deferred questions to legal officials, who did not return calls by late Tuesday.

Since learning of the charges July 8, Keeler said he has not been appointed a military attorney, and that retaining a civilian attorney costs more money than he can afford.

The 23-year-old Marine was deployed to Iraq serving as a sniper with the 3rd Battalion, 5th Marine Regiment. When he returned to Camp Pendleton along with about 200 other Marines, he said he was told to put his gear in an empty room before taking four days of liberty.

Keeler said he drove up to Ventura to see his younger brother during his time offduty. Once he returned to Camp Pendleton on May 21, he found that his gear had been moved and that some of his equipment, including some night vision goggles, was missing.

Hunting for his gear, he heard privates say that an Iraqi-made pistol had been found. Keeler said he decided to try to recover the weapon to return it to the proper authorities and keep it away from the younger Marines.

In doing so, he told the privates that the pistol was his.

The pistol was found in the vehicle of a Marine private, but because Keeler had claimed it was his, he was charged with the alleged crime, he said.

"I was just trying to recover it," Keeler said. "I believe that higher-up officers in the Marine Corps are looking for a scapegoat, trying to make an example out of me and trying to unjustly charge me with a crime I did not commit."

Keeler was subsequently charged, under the Uniformed Code of Military Justice, with disobeying an order by "wrongfully retaining and introducing (the weapon) into the U.S.," according to a military document signed by Capt. B.J. Cooke, a Camp Pendleton judge advocate.

The document also says Keeler made false statements claiming that the pistol was a "family heirloom" that "had been (in) his family for about 20 years." Keeler said he never made such statements.

A Washington state native, Keeler was looking forward to leaving the military in March to attend college next spring. But because of the war with Iraq, the Marine Corps placed an order keeping troops from leaving the service.

In Iraq, Keeler said he served in night patrols in Baghdad. During one of those patrols, a fellow Marine, Lance Cpl. David Owens Jr., 20, was shot and killed while taking a defensive position next to Keeler.

Because of the charges, the young sergeant said he fears his otherwise exemplary military career is now tarnished.

"Regardless of what happens from this case, it's already drastically, negatively affected my life," Keeler said.

Contact staff writer Edward Sifuentes at (760) 740-5426 or esifuentes@nctimes.com.

Though the above article had a lot to say about the Iraqi 9 mm pistol, it had nothing at all to report about the fate of the stolen goggles or who had absconded with them in the first place. It seemed clear that Mr. Keeler, like Dion, was most likely *not* the culprit.

* * *

By December I was almost as much of a wreck as Dion because I had to keep taking his calls while trying not to get lured into the quagmire of Pacific Beach. I was teaching several classes at CSU Long Beach from August to December, so I could always use that as an excuse as to why I couldn't go down and observe all this strangeness firsthand. But then at the beginning of January I wasn't assigned any classes to teach for the spring semester. This was worrisome, not because I had to go on unemployment but because now I had no good excuse as to why I couldn't go south and save Dion from the horde of invisible midgets surrounding his apartment. Dion made the situation even worse by trying to convince me that the reason I didn't receive any classes was because the U.S. government had called the chair of the English Department and convinced her I was a known terrorist and a security risk to the United States of America. I began to wonder if this could be true.

I tried to find whatever information I could through the network of contacts I had made while writing for *Paranoia* magazine. I contacted Walter Bowart, author of *Operation Mind Control*; Jon Rappoport, editor of *U.S. Government Mind Control Experiments on Children* (who lives in San Diego); a little known metaphysician and psychic researcher named David Worcester who lives in Seattle, Washington; and other experts in government mind control programs.

Walter Bowart was a font of valuable information. He had been researching mind control for almost thirty years. I knew him fairly well for several years before his death from colon cancer in December of 2007. We even collaborated on a feature-length screenplay entitled *The Other Crusades*, which was all about Walter's adventures in the 1960s trying to prevent LSD from being made illegal by the U.S. Congress.

I believe it was Walter who directed me to a website called "Raven1.net" where I discovered a brief article entitled "All About Street Theatre"[4] by Eleanor White. The

---

4. Until recently, Eleanor White's article could be found at the following link: http://www.raven1.net/abtstth.htm. White's website has since disappeared, but you can see what it once looked like at this link: web.archive.org/web/20060927223449/www.raven1.net

article described, in exact detail, almost everything Dion had undergone during the past few months.

ALL ABOUT STREET THEATER

"Street Theater" is a feature of the organized stalking with electronic harassment scene.

"Street theater" is activity performed by persons complicit in the electronic weapons harassment, but are "skits," as opposed to direct bodily attacks performed with the electronic harassment equipment.

These "skits" are designed to imitate "the breaks" of normal living. Additionally, they are performed in such a way that the target, and ONLY the target, knows they are being harassed, but cannot convey to others that this is indeed harassment. Feelings of total hopelessness, and that "everyone is against me" is one apparent purpose of these "skits."

(What is impossible to convey to people who are not targeted is that what is different about mind weapon research skits is QUANTITY. When you encounter "normal breaks of life" several times a day EVERY DAY, you are no longer talking about "normal." Several "breaks" a day, of a type which you might expect every couple of months, is not natural or random. But try explaining this to someone who is not targeted.)

Another apparent purpose of such "skits" is to discredit and isolate the target so that others will regard him or her as a "crank" and a "nut case" when the target complains.

Far from simple "pranks" or "practical jokes," these skits provide the mind weapon researchers with extremely good cover. If the target is ever coerced into contact with psychiatry, the psychiatrists' legal powers of imprisonment (without due process of law) dramatically increase and reinforce the isolation and labelling of the target.

Many people know in advance that what they are experiencing will discredit them, and will thus put off complaining about or often, even admitting to themselves that they are being targeted.

So although "street theater" seems to have a comedic ring to it, this component of organized stalking is one of the most serious forms of attack on individual targets and is perfect cover for the perpetrators […].

Street theater takes many diverse forms, and here are a few examples:

- On foot, far more often than in normal life, you have people cutting you off in store or bank lineups. Or you constantly find people getting "in your face" as you walk, both outdoors and especially in buildings and malls.

- While driving, far more often than in normal life, you have cars speeding up to stop signs just ahead of you and brake to a stop part way into the intersection.

- While driving, far more often than in normal life, you find other cars behaving in ways which block your progress. Mall parking lots are favourite places for this type of staged activity. Try explaining that to friends and see how many believe it is deliberate.

- While away from home, more often than in normal life and at times you know you did not leave them, dirt or food droppings appear in your house or apartment.

- While away from home or work, belongings turn up missing and you know for certain they were there when you left. Some days later these belongings may turn up in a place you know they were not, yet you cannot ever convince others this was theft and return.

- While away from home, you find damage to clothing or furniture which you know did not occur from normal wear.
- While at home or at work, you find bizarre, loud, annoying noise incidents which others nearby seem to "not notice" or "don't care about."
- While in the supermarket checkout line, you find someone reaching into your shopping cart to remove an item—apparent purpose to force you to make another trip to the store prematurely.

These are just some examples of "street theater." The number of variations on this "wear-you-down" activity seems unlimited, based on reports.

This article encouraged me to keep digging for more information on the subject. I learned that there were various labels for what Dion was experiencing, "street theatre" and "gang stalking" being two of the most common. This was a phenomenon by no means limited to San Diego. After the events of 9/11, these acts of domestic terrorism on innocent civilians increased a thousandfold. For some of the victims the reason was obvious (as in Dion's case), while in other instances the impetus for the stalking seemed random.

✳ ✳ ✳

By December Dion was barely able to concentrate on anything except living from second to second. He sent me a photo of what he looked like by this point. It was as if he had become a different person. All the life had been sucked out of him in an eerie, almost supernatural way. I'd witnessed the effects of meth and heroin and general madness on plenty of different people before this point, but this was like nothing I'd encountered.

Dion had lost so much weight that he seemed to be little more than a hollow husk, a walking mummy that might crumble into dust at the merest touch. Quite frankly, he looked more like photos I'd seen of people suffering from slow radiation poisoning. He told me he often had a metallic taste in his mouth these days. From my discussions with Walter Bowart, I learned this was a common side effect of intense bombardments of electromagnetic radiation.

Lately Dion had been complaining about high-pitched noises that would emerge from nowhere in the middle of the night, inducing migraines and vomiting. What were these people trying to do to him? Break him down or kill him? Perhaps a little of both. By this point I think Lita and her goons were so stunned that he was still alive that they just wanted to see how much more he could take before he escaped the maze or died.

Walter Bowart also turned me on to a corporation called American Technology Corp. located at 13114 Evening Creek Dr. S. in (where else?) San Diego. They specialized in "High Intensity Directed Acoustics." According to Marshall Sella of the *New York Times* (3/23/03):

> [Woody] Norris and A.T.C. have been busy honing something called High Intensity Directed Acoustics (HIDA, in house jargon). It is directional sound—an offshoot of HSS—but one that never, ever transmits Handel or waterfall sounds. Although the technology thus far has been routinely referred to as a "nonlethal weapon," the Pentagon now prefers to stress the friendlier-sounding "hailing intruders" function.
>
> In reality, HIDA is both warning and weapon. If used from a battleship, it can ward off stray crafts at 500 yards with a pinpointed verbal warning. Should the offending vessel continue to within 200 yards, the stern warnings

are replaced by 120-decibel sounds that are as physically disabling as shrapnel. Certain noises, projected at the right pitch, can incapacitate even a stone-deaf terrorist; the bones in your head are brutalized by a tone's full effect whether you're clutching the sides of your skull in agony or not.

"Besides," Norris says, laughing darkly, "grabbing your ears is as good as a pair of handcuffs."

Nimbly holding a big black plate, Norris stands with me in an A.T.C. sound chamber. Since he's poised behind the weapon, he will hear no sound once it's powered up: not a peep. "HIDA can instantaneously cause loss of equilibrium, vomiting, migraines—really, we can pretty much pick our ailment," he says brightly. "We've delivered a couple dozen units so far, but will have a lot more out by June. They're talking millions!"

(Last month, A.T.C. cut a five-year, multimillion-dollar licensing agreement with General Dynamics, one of the giants of the military-industrial complex.)

Norris prods his assistant to locate the baby noise on a laptop, then aims the device at me. At first, the noise is dreadful—just primally wrong—but not unbearable. I repeatedly tell Norris to crank it up (trying to approximate battle-strength volume, without the nausea), until the noise isn't so much a noise as an assault on my nervous system. I nearly fall down and, for some reason, my eyes hurt. When I bravely ask how high they'd turned the dial, Norris laughs uproariously. "That was nothing!" he bellows.

"That was about 1 percent of what an enemy would get. One percent!" Two hours later, I can still feel the ache in the back of my head.

Note: The San Diego headquarters of A.T.C. is located only sixteen miles from Dion's Pacific Beach apartment, a mere twenty minute drive.

7.

Meanwhile, I was spending my days trying to find a job to tide me over until the fall semester started up. My paranoid state wasn't helping matters. A Bookstar (now defunct) located in a mini-mall on Hawthorne Boulevard in Torrance rejected me because I had a Master's Degree. "If I hire you, you're gonna steal my job!" the manager said in a rare moment of candor. I was rather surprised when she literally laughed in my face as I handed her the application. I left the store ashamed and angry, wondering if the invisible midgets were to blame for my misfortune.

At around this time Dion called to tell me he had seen some weird phenomena near the pier. The psychological terrorism had grown so bad that Dion fled the house and went to the beach, hoping the open air would do him some good. That's when he saw the tracks in the sand. Some kind of vehicle, he claimed, followed him along the beach. He saw its tracks forming in the sand, but couldn't see the vehicle itself. He ran from them, only to see the tracks speeding up behind him. He even saw their tracks overlapping *his own footprints*. He led these invisible vehicles in circles around the beach for about twenty minutes, then climbed up onto the pier and dashed back home.

The next night at around three in the morning he took a friend, some skinhead meth addict, back to the beach to show him what had happened. The place was deserted. Dion hoped for a repeat just to be able to prove it to someone else. They didn't encounter the invisible vehicle, but saw something even better. They saw what Dion described as a dozen or so robots marching down the beach, in jerky, stop-start motions like a Dynamation figure from a 1950s Ray Harryhausen movie. Dion said he wasn't sure how to describe it, but it looked like the things were leapfrogging *sideways*. For a moment it seemed as if the creatures were going to topple over, then they would suddenly leap up about fifteen to twenty feet in the air. These things appeared quite creepy marching along the moonlit beach, so Dion and his friend got the hell out of there quick.

This was another aspect of Dion's Munchausen-like tales that I found almost impossible to accept... until I saw "BigDog" and "Petman" in action. If you'd like to see video of the experimental military robots Dion and his friend undoubtedly witnessed that evening, check out the following links to a pair of 2009 news reports about "hopping 'bots" that might "one day save lives on the battlefield": 1) "Video: Meet Petman"[5] and 2) "Big Dog Maker Tests Humanoid 'Petman' Robot, Hopping 'Bot."[6] The latter article, dated 4-20-09, reads as follows:

> Boston Dynamics Inc., maker of the YouTube sensation BigDog robot, reports it is developing two robots: a humanoid robot to test soldier uniforms and a hopping robot for navigating urban environments.
>
> The humanoid robot, PETMAN (protection ensemble test mannequin), is planned to be used for high-fidelity testing of chemical protection suits under chemical warfare agent exposure conditions. The robot is intended to balance itself and walk, aim, crawl and perform suit-stressing calisthenics. The robot will also mimic human physiology with the protective suit, simulating the act of sweating, for example.
>
> The PETMAN robot is a $26.3 million U.S. Army project. Boston Dynamics is working with the Midwest

---

5.   http://video.foxnews.com/v/3943665

6.   http://www.masshightech.com/stories/2009/04/20/daily6-Big-Dog-maker-tests-humanoid-Petman-robot-hopping-bot-.html

Research Institute, Measurement Technology Northwest, Smith Carter USA and HHI Corp. on the robot, which is expected to be delivered in the second quarter of 2011.

The Waltham Robotics Company also reports it has landed a deal from the Sandia National Laboratories to develop the Precision Urban Hopper, a four-wheeled robot with one leg to facilitate jumping over obstacles. The robot is intended to navigate autonomously and jump 25 feet in the air.

Development of the robot is funded by the Defense Advanced Research Projects Agency. Financial terms of the deal were not released.

Dion said this was all so damn strange he almost wanted to believe he hadn't seen the robotic creatures. But this time there was a witness.

And several of Dion's junkie acquaintances made it clear to him that they didn't come over anymore because of all the insane shit that was happening at his apartment. They always felt like they were on the verge of being arrested at any moment, as if someone were watching their every move.

The strangeness increased, as strangeness had a tendency to do around Dion. About a week after the hopping robot incident some other junkie acquaintances, Adam (AKA "Adam Splat'em") and his childhood friend Chris, returned from a trip in a nearby desert area. One afternoon Dion ran into Adam on Garnet Avenue. A little reluctantly, Dion told him about the leapfrogging 'bots. Instead of telling him he was crazy, Adam said, "Dude, let me tell you *this*. Me and my buddy were drivin' this dune buggy all around the desert.[7] We were in the middle of fuckin' nowhere, right? Well, at around noon we come across these people. A bunch of 'em. We could see them from

---

7.   Somewhere in (or very near) Santee, CA.

far away. Just black dots off in the distance. They were all lined up in a row. As we got closer we could see that they were firing guns. It looked like they had these targets set up and they were shooting at them. I swear to God, we weren't drunk, man. We weren't high out of our minds. Sure, we were drinking a few beers, but so the fuck what, right? We thought these guys were maybe just some hippie desert rat types target practicing. You know what I mean? Anyway, we figured we'd swing by and get a closer look. At first we didn't think we were seeing what we were seeing, but... I swear to fucking God, man: you can ask my buddy, he saw it too. These guys, they weren't human. They were fucking *apes*, man. They were standing upright and they were dressed in uniforms... like, military-type uniforms. A couple of them looked at us, and their eyes were fuckin' scary. They were *intelligent*, man. So we turned around and got the fuck out of there as fast as we could. They went back to their target practice and didn't fuck with us, thank God."

Dion asked him if they looked like humans wearing ape suits, or if they really looked like apes. He said, "No, man, you can tell the difference between some dude wearing a suit and a real living *being*. These things were apes and they were standing upright and they were breathing and aiming guns and fuckin' *staring* at us, man, like they were going to say something to us, but they didn't because we got the fuck out of there." He added that the apes were (and this is an exact quote) "huge, silent, graceful."

Dion asked if he remembered whether or not there were humans mixed in with the apes, and Adam said, "No way, man." Then he thought about it and said, "You know what, dude, I don't know. I don't think so. All I saw were those fucking apes. I didn't see any humans around at all. Maybe there were, but I didn't see them."

Dion came to the conclusion that there was some connection between the desert apes and the robots. If so, I'm not sure what it would be, except for the fact that there's a heavy military presence in both Pacific Beach and the desert area where Adam and his buddy were gallivanting around. That particular desert is no-man's land. If you wanted to conduct some freaky covert experiments out there, like letting your genetically-engineered Moreau-type human/ape hybrids climb up out of your hi-tech underground military base for a Sunday afternoon stroll and some leisurely target practice on cardboard figures shaped like Iraqi terrorists, then that would probably be the place you'd want to do it.

8.

As I typed the last sentence (on June 3, 2010), I heard a report on the radio that the percentage of crimes solved by the local police in San Diego is a whopping 94 percent. In L.A. the percentage of crimes solved is around 40 percent. In Chicago, it's even less. The asshole conservative talk show host who was talking about this then said, "Wow, San Diego's really got it goin' on down there. What the heck are they doin'? What's their secret? Whatever it is, they should be teachin' it to the LAPD."

I thought, Jesus, I'm not surprised 94 percent of the cases are solved, particularly when you've got an army of invisible midgets, leapfrogging robots, and gun-toting genetically-engineered simians running around terrorizing everybody in the city. "By God, if we can't find the person responsible for this heinous crime, I pledge that we will find the *next best suspect* and drive them over the deep end until they confess to anything and everything!" I think the remaining 6 percent are those people who died of fright before they were able to confess. Or those who somehow managed to escape The Village.

Not long after being told the ape story, my friend Eric Williams and I were invited to attend a screening at The Midnight

Special Bookstore (now defunct) on the 3rd Street Promenade in Santa Monica. The man who invited us to the screening was Gerry Fialka, founder of M.E.S.S. (Media Ecology Super Sessions), a proletarian think tank for fringe artists and freakazoids of all kinds. The screening was for a documentary called *Cul de Sac: A Suburban War Story* directed by Garrett Scott.[8] This documentary is the story of a man named Shawn Nelson, a thirty-five-year-old ex–Army soldier and resident of Clairemont, a suburb of San Diego next door to Pacific Beach—essentially the *same neighborhood* in which Dion lived. In May of 1995, Nelson stole an entire tank from the local National Guard Armory and drove it down the middle of San Diego for twenty-three minutes, decimating fire hydrants, lampposts, automobiles, RVs, houses, and freeways, causing thousands upon thousands of dollars' worth of damage. It was miraculous no lives were lost that day.[9]

Wait, that's wrong. Someone *was* killed that day: Shawn Nelson. At one point, on his way toward City Hall, the tank got stuck on a concrete median right in the middle of State Route 163. Unable to extricate the tank from this situation, the cops surrounded him.

It's incredible to watch this footage of mass destruction. I think it resonates with a small part in all of us, that part that

---

8. It's worth noting that Garrett Scott died unexpectedly at the age of 37 in March of 2006 while swimming in a municipal pool in Coronado, CA (located in San Diego County). According to the 3-10-06 edition of the *New York Times*, Scott passed away "two days before his documentary *Occupation: Dreamland* received an Independent Spirit Award at a ceremony in Santa Monica […]. His death was announced at the awards ceremony on Saturday by his partner and co-director, Ian Olds, who accepted the award for both of them […]. *Operation: Dreamland* was honored with the organization's Truer Than Fiction Award for its account of the weeks in 2004 that Mr. Scott and Mr. Olds spent embedded with the 82nd Airborne Division of the United States Army. They were allowed to shadow eight members of the unit deployed in the Iraqi town of Falluja." The official cause of Scott's death was a heart attack.

9. Upon seeing this documentary, I felt silly for wondering how the hell Lee/Doyle had stolen a mere *truck* from Camp Pendleton.

sometimes grows so damn frustrated you just want to FUCK SHIT UP. It's the ten-year-old in all of us who responds so favorably to the rampant chaos in *Godzilla* and similar films.

The cops weren't amused. You can see about four of the police officers literally pushing each other out of the way for the opportunity to open the tank and shoot this guy in the neck. Which is what they did, despite the fact that Nelson was unarmed (except for the now-useless tank). Officer Paul Paxton, a Gunnery Sergeant with Alpha Company, 4th Tank Battalion, a Marine Corps Reserve unit in San Diego, opened the hatch with a pair of bolt cutters. Paxton's partner, Richard Piner, pulled the trigger. Nelson died instantly.

This is one of those dramatic incidents that, once it's over, slides away into the memory well and no one mentions it ever again—unless some clever fellow, like Garrett Scott, decides to film a documentary about the incident. It was an incredible coincidence that Fialka should have been screening this documentary at that exact time. It only confirmed every suspicion I'd had over the course of the past seven months: the crazy and inescapable suspicion that Dion wasn't hallucinating.

As the documentary unfolded, I recognized Dion's neighborhood. I recognized the types of people interviewed in the documentary, the "tweeker" milieu so inextricably linked with that particular speed-wracked corner of San Diego. Many of these people knew Shawn Nelson personally. They were all meth addicts. And it was clear that the local police had them under constant surveillance. One cop, George Eliseo of the SDPD, drives through the neighborhood and points out every single house in which a drug user lives. He implies that he knows the location of each "parolee house" and that the department has them under surveillance at all times. He refers to these people not as criminals or addicts or even humans of any kind, but as "bottom feeders."

The filmmakers interview Nelson's close friends, all of whom describe him as intelligent and not crazy in any way. One friend said that Nelson believed "he was being harassed by the city." His brother, Scott Nelson, said, "He'd had enough. He was tired. He was trying to get some attention." All of them said that Nelson had dug a seventeen-foot-deep hole in his backyard. He convinced all his meth-addicted neighbors that there was gold buried in the yard in order to rope them into helping him dig the hole. I suspect this may have been a clever ruse on Nelson's part. If you're being bombarded by sophisticated electronic warfare devices, it might be a good idea to dig a seventeen-foot-hole in your backyard. Such a deep hole might act as a kind of makeshift Faraday cage to lessen the impact of the microwaves and/or electromagnetic energy being deployed against him. This, of course, is just speculation, a possibility that passed through both my mind and Eric's as the documentary unfolded before us.

Nelson's former employer, Dale Fletcher, offers this intriguing peek into Nelson's psyche:

FLETCHER: Shawn was adamant that there was a government helicopter that was chasing him, that was trying to shoot him, and it was like, you know, I couldn't… I couldn't really believe it, you know? If that really happened, my goodness… you know?

INTERVIEWER: Would you just listen to him or would you try to reason with him?

FLETCHER: No, not at all try to reason with him. I mean, he was—this is what *happened* to him.

One man named Chuck Childers, Nelson's best friend during the last year-and-a-half of his life, had an even more chilling story. Childers claimed Nelson's ultimate goal was to drive the tank up onto the steps of City Hall so he could make a state-ment in front of the media. Childers' tales of harassment and psychic terrorism in the Clairemont/Pacific Beach area mir-rored the exact horror stories that Dion had been telling me since July. At one point Eric and I, sitting in the back of the small screening room, just looked at each other silently, our thoughts so clear they need not be spoken.

In the documentary Mr. Childers gives intriguing testi-mony about the people who had been spying on Shawn Nelson for months and no doubt playing psychological warfare games with him. It's clear from the way Garrett Scott has chosen to edit this footage together that he doesn't take what Childers is saying very seriously. I suppose most sane people wouldn't. However, when juxtaposed with Dion's experiences in the exact same neighborhood, Childers' peripatetic "tall tales" take on a far grander significance...

CHILDERS: You know what a subliminal is, right? This par-ticular optical-digital technology is very capable of pro-ducing the same kind of thing. It's being pulsed on a laser, but your mind, your subconscious, is digital. It picks it up. So it can literally be programming you. Subliminally.

INTERVIEWER: And you wouldn't know it.

CHILDERS: Right. One day you wake up and all of a sud-den our morals that used to be over here are over *here* and you don't even know it—that they've changed. Because it's been a gradual thing. When I read that in the library it kind of spooked me.

INTERVIEWER: Now this brings us back to Shawn a little bit. A lot of these hallucinations you were talking about. Do you think they came from this—?

CHILDERS: Well, I don't know that he had hallucinations because he never said specifically that he saw things or people or whatever. But stuff that *happened* to him. Like

I said, he'd be discussing something with this person over here, and he'd go across town and they'd be talking about the same thing. Maybe not *to* him, but the people next to him would be talking to each other about the same identical... almost word for word... you know. And at first you thought, "Oh, you know, coincidence," but then it kept happening and happening...

Nelson would have a conversation with someone, then two hours later on the other side of town hear strangers echoing the same exact conversation back at him. This must be a common technique these perps employ, designed to confuse and unbalance targets, as Dion reported these same incidents to me over and over again. It's a form of Chinese water torture. Perhaps it doesn't sound so maddening... until you experience it yourself. If someone said, "I'll pay you five hundred bucks and all you have to do is sit here and allow me to drop little pellets of water on your forehead for a couple of hours," you might be tempted to accept that offer. Until about ten minutes later when you're begging them to take five hundred dollars out of your wallet just to make it stop. These techniques are simple but effective. If you heard groups of people talking about you everywhere you went, and the comments were repeated throughout the day from completely different groups who didn't seem to know each other, you would begin to grow so paranoid that you'd start trying to crawl out of your own damn skin. Think about how self-conscious you get when you hear a mere acquaintance criticizing you for some stupid, trifling little thing. Magnify that mortification a few dozen times, from disparate groups of people you've never met before—and, on top of that, whenever you confront these complete strangers, they ignore you as if you don't exist, or deny that they were ever talking about you in the first place. It wouldn't take long for most people to lose it. Now add a steady stream of meth, invisible midgets, and sharpshooter ape-men into the mix and just try to imagine how long you'd survive the onslaught. Not even the *Mission: Impossible* team ever fucked with a Central American dictator as much as these jarheads fucked with Shawn Nelson and Dion Fuller.

I should point out that the people following Dion did not fit one particular description. Sure, some of them looked like military, but a lot of them didn't. Some of them looked like soccer moms. Real normal people. It's not clear to me how these people are recruited. It's possible they have no idea who Dion was or what he'd supposedly done. Perhaps they were given orders and followed them without question. Perhaps they were told that Dion was a suspected terrorist. Which, of course, would be a true statement. He *was* a suspected terrorist— suspected by the NCIS, who made up the charge in the first place. A mere two years after 9/11, it would only have taken a charge like this to raise the ire of any red-blooded American. P.T. Barnum was right. And Adolf Hitler was also right when he wrote in *Mein Kampf*, "…in the big lie there is always a certain force of credibility…"

As I said before, the testimony of the people featured in *Cul de Sac* was quite stunning. What was even more stunning was the fact that it was clear, from the way the filmmakers juxtaposed certain scenes, that you were supposed to find these "ravings" to be amusing. The filmmakers thought they were making a movie about the economic disintegration of San Diego in the wake of the Cold War, but in fact it was vital documentation regarding nefarious, widespread government corruption, a phenomenon that Garrett Scott *accidentally* stumbled across and captured on video. The filmmakers themselves had absolutely no idea what their movie was really about. Which means you can be in the thick of this gang stalking milieu and not even realize it… unless, of course, *you're* the one being harassed. It's the perfect cover. Edgar Allan Poe, in his famous short story "The Purloined Letter," recognized as far back as 1844 that the most effective hiding place is right out in the open where everyone can see you.

Finally, *Cul de Sac* offers yet one more revelation of staggering proportions: If this sort of experimentation was already in high gear back in 1995, then obviously the stolen goggles were not the cause of all this jabberwocky. The ultimate purpose of this project is far more overreaching than just National Security concerns. Perhaps the situation is best summed up by Chuck Childers: "We're just supposed to be part of The Machine. Kind of like what Hitler was trying to do. What Stalin tried to do. Communism. That's exactly what it is. *1984*. George Orwell. It's goin' on right now. Big Brother's watchin', probably right now as we speak. And controlling. Or trying to. They're doing a very good job of it. They haven't got total control yet."

If you stand out in any way, if you do anything just slightly out of the norm to bring yourself to the attention of these psychopathic stormtroopers, the jackboot comes hurtling down from the sky and stomps on you repeatedly until you either agree to merge with The Machine or you die. As Orwell once wrote, "The future is a boot smashing a human face forever."

Right before all the nonsense with Lee even started, Dion spent some of his evenings wandering all over Pacific Beach hanging up inflammatory, pornographic, anti-Bush flyers in the form of crazy collages that he would hang on telephone poles and pay phones and windshields of parked cars and prominent store fronts. Is *that* what originally caused him to come to their attention?

The night vision goggles were the trigger, of course, but it's clear to me that the stormtroopers already had this operation in place in San Diego for years. That's why their response was so immediate. The second Dion was released from jail, the stalkers were all over his ass. The chess pieces were already in place. The numbnuts in charge just move the pieces around at their discretion whenever anyone they don't like crosses their field of vision. This is probably why so many of these gang stalking victims have absolutely no idea why the harassment started in the first place.

Maybe you cut off Lita Johnston on the freeway. Then, with the push of a button and a signature on the right form, you're now a suspected terrorist. And your life has been turned upside down until you either kill yourself or steal a tank and let someone else do it for you.

This couldn't be allowed to happen to Dion. He wasn't going to be able to beat these fuckers at their own game, so he'd have to up and leave San Diego. But how could he escape?

9.

In a moment of desperation, during the middle of a random weekday, Dion came up with the idea of climbing onto the roof of his apartment building in order to get a better view of the situation. This must have discombobulated the perps, as he was able to see the entire pattern of the operation, including the perps' mobile command center; all the surveillance activity radiated outward from a large windowless white truck parked surreptitiously a few blocks away. From this new perspective, it was much easier to separate the stalkers from the civilians. The stalkers were the ones now staring up at him in surprise. Some even pointed at him and talked heatedly among themselves, as if panicked, unsure of what to do now that they had been spotted. For a brief moment, for the first time since July, Dion felt like he had a little power, a modicum of control over his own destiny. But since he couldn't stay on the roof forever, his godhood didn't last for very long. It did offer him a tiny slice of hope, however, no matter how short lived.

10.

Dion called me during the first week of January 2004 and told me he'd run into some homeless meth addict who was looking to sell his van for a song, only seven hundred dollars. He was almost willing to *give* it away. For some reason the guy just wanted to get it off his hands. Perhaps there was a dead body in the back. Who knows? The dude claimed he was going back home to Chicago to live with his family and wouldn't need the van anymore.

Well, it just so happened that Dion had exactly seven hundred dollars (thanks to two unemployment checks and a check that his dad had sent him a few days before). Dion had these three checks in his wallet. He was all prepared to buy the car when he realized his Driver's License was no longer in his wallet. To make matters worse, the next day the *entire wallet* disappeared out of Dion's living room. So now he had to wait for his next unemployment check before he could buy the van. That wouldn't be for another two weeks. The meth addict told Dion he'd have to sell it to someone else if Dion couldn't come up with the cash in a couple of days.

Over the phone I could hear the desperation in Dion's voice. He was hanging on to his sanity by the merest filament. The mindbending scenery shifts outside his window had grown more ominous. He could barely describe them to me as they were happening. At one point I was talking to him on the phone and he cut off what I was saying and described, with some puzzlement, some kind of red light beaming through his window. A second later he yelped in fear as a plate flew off the shelf and landed on the ground in front of him. The window next to his head shattered into pieces. Dion has a macho self-image he tries very hard to protect at all costs, *and* he's a horrible actor. That's why I don't think he was putting on some elaborate hoax for my benefit when I heard him scream as the window shattered next to his skull. He fell flat on the ground and was whispering into the phone that it was all over. Now they were taking shots at him. They were going to *kill* him… or at least that's what he thought at that moment. It took him a good twenty minutes to calm down. Nothing else happened for the rest of the night, but I could tell Dion was on the edge of losing his shit at last. And it's not that I believed Dion didn't deserve to lose his shit after some of the weird stunts he'd pulled in his life, but it bugged me to think these pervy black pajama boys were going to get the last laugh in this situation.

Dion insisted they were somehow turning off all the noise in his apartment. "They're doing the silence thing again," Dion would say. And I could *hear* this through the phone. Suddenly, all the ambient noise would just… cease to exist. It was as if the Cone of Silence machine from that old *Get Smart* TV show had dropped down around him. Total silence may not sound so torturous, but if you experienced it I think you'd feel differently. Again, a lot of these techniques were tantamount to Chinese water torture. For a few minutes it might not be so bad, but for a few hours? Days? Months?

One night later, a two-and-half-foot amoeba-like creature crawled across his bedroom floor, crept up the side of a chair, and slithered into his leather jacket which was sprawled on the floor. The jacket filled out with the amoeba's presence and began inching across the carpet. Dion watched the jacket creep toward him for a few minutes until it deflated and lay still, the amoeba nowhere to be seen.

Bike parts in the corner of the room began to twist and turn, as if they were either trying to kill each other or have sex.

Ankle-sized gnomes danced across the carpet, then vanished.

Just after the New Year, while having dinner at the charming but cramped Venice home of Gerry Fialka (the aforementioned founder of M.E.S.S.), Fialka's wife Suzy introduced me to her friend Joan Sienkiewicz, a Santa Monica attorney. I told her about Dion's plight in detail. Since she'd read about directed energies harassment cases before (though perhaps not to such a surreal degree), she suggested I contact Harriet McDougal, an attorney with experience representing survivors of organized ritual abuse and torture. Joan also gave me the email address of Cheryl Welsh, founder of Citizens Against Human Rights Abuse, AKA Mind Justice (a non-profit organization that Welsh describes as "a human rights group working for the rights and protections of mental integrity and freedom from new technologies and weapons which target the mind and nervous system").

On January 3, 2004, I emailed Ms. Welsh a brief rendition of the whole story. Only a few hours later, Ms. Welsh wrote me back:

Dear Robert,

Yes, I have no trouble believing your account. The technology to harass is commonplace and widely reported by victims. Off the record it exists but officially no. My group is small, research and networking.

I would recommend a private investigator who knows about intel ops as your best bet, my best advice. Sure it takes money, but I know of no other way to have a chance at stopping this. Well, there are many opinions out there, take this one or leave it! I do use a world class private investigation company in SF, that knows about this technology, so that first hurdle, that you aren't crazy, is behind you. It's Lipset Investigations, Kyle is who I have worked with. Well, whatever you do, best of luck and if I can help further, let me know.

—Cheryl

I was grateful for the encouragement, of course, but neither I nor Dion could afford to hire a private investigator. I thanked Ms. Welsh for her advice and moved on to the next plan. Unfortunately for Dion, there wasn't a "next plan"—not yet, at any rate.

Another incident occurred during the second week of January. At around noon one day Dion sat down to eat lunch, a deli sandwich he'd ordered from a take-out place down the street. Only seconds after taking a few bites he grew ill and felt suddenly fatigued. He lay down on the couch. He didn't remember falling asleep, but he did remember the dream when he woke up hours later after dark.

The dream was this: he was lying on the couch, unable to move, as several men entered his apartment through the back door and crowded around him. One of them kneeled down beside him and pulled out a hypodermic needle. He remembered trying to scream and move, but couldn't. It was the sort of nightmare one might have when experiencing sleep paralysis. Except for the fact that this didn't feel like a "dream" at all. He recalled the fear as he felt the needle being slipped into his arm. He didn't remember anything after that except waking up still feeling as if he hadn't slept for days.

Based on his vivid description, I was almost certain this was no dream. Dion never experienced nightmares like that. The types of dreams that plagued Dion's sleep were those in which he was stuck behind a cash register for eight hours a day or endlessly doing loads of laundry. I think mundanity frightened him more than anything else. This was the only time during the fifteen years I had known Dion in which he associated a needle with *fear*. I wondered what the hell these people were planning for him. It felt to me like the situation was growing worse and worse, as if the perps were growing more brazen.

Dion never asked me for any money, and no doubt thought I didn't have any to give. I didn't tell him what I was going to do, as I didn't want the transfer of funds to be intercepted in any way. I knew that everything I said to Dion on the phone was going into a very entertaining file that, alas, no one would ever read. But I don't need to initiate a FOIA request to get ahold of it because the version I'm writing now is probably far more accurate. I take good notes. I always have. Remember that.

I contacted my friend Eric and asked him to meet me in the parking lot outside my apartment building in Torrance. He picked me up at around ten o'clock at night. I directed him to a Bank of America. On the way there I explained to him the latest situation and told him I was going to pull seven hundred dollars (which I could ill-afford) out of my savings account and send it via Western Union first thing in the morning so Dion could immediately pick up the money and buy the van and get the fuck out of San Diego for good. Eric said, "Jesus, you're a good friend." Without me asking him, Eric volunteered to give me two hundred and fifty so I didn't have to sacrifice the entire seven hundred dollars. He made it clear, however, that I shouldn't mention his name over the phone.

The next morning, around 8:00 A.M., I strolled down to the Western Union in the back of the Fox drugstore on El Prado Avenue in downtown Torrance. I gave the young woman behind the booth the seven hundred dollars and Dion's name. She told me the funds would be available for pick-up within the hour at a Vons only a few blocks away from Dion's apartment. I strolled back home, glancing furtively over my shoulder, wondering if the homeless people lurking in the alleys were really homeless people, or were they something else altogether? Were they even human? Were they apes that had been altered to seem like humans? Who were they working for? And what about the spies I *couldn't* see?

I got home, locked the door behind me, entered my bedroom, and picked up the phone. Only a few nights before, I had been calmly reading a book when a loud beeping noise started going off in my room. It was unlike anything I had ever heard before (it wasn't debilitating, just annoying). I was never able to trace the source, though I spent twenty minutes trying to do so. It wasn't an alarm clock, nor was it a fire alarm or a cell phone. I owned no equipment that made a high-pitched noise like that. I'd never heard a sound like it. Some experts say listening devices are so sophisticated that you would never be aware of their existence. That may or may not be true. Everyone fucks up from time to time, even invisible homunculi.

I called Dion at around 9:00 A.M. "Yeah?" he said, sounding deflated.

"How're you doin'?" I said.

His laughter was laced with hopelessness. "I really thought they were going to kill me last night."

"Is that guy still offering you his van?"

"Yeah, but I wasn't able to scrape up the money."

"I just wired seven hundred dollars to you. It should be available in half an hour."

He was silent for a moment. "Holy shit, are you kidding?"

"No. It's there. It's waiting for you."

"Fuck, *thank* you. Jesus, I owe you big time."

"It's okay. Do you want the password?" In order to pick up cash from a Western Union office, Dion would need a photo I.D. Since he no longer had one, that meant I was required to make up a password that Dion would have to provide to Western Union before they would give him the money.

"No, don't give it to me now," Dion said. "I don't want anything to fuck this up. If they know the password they could send one of the midgets down there to pick up the cash before I can even get to it. I have to go call the guy right now and tell him to hold onto the van for me. When I'm ready to walk over there you can give me the password then."

That sounded reasonable. "Okay," I said.

About an hour later he called back and said, "I couldn't get him on the phone so I just walked over to his place. Okay, I'm ready to go to the store now. What's the password?"

I said, "The password's—" At that moment the line went dead. "Hello?" I said. Silence. "Dion?" Then a dial tone.

I called back. It rang twice. Dion picked up. "Hello?" he said.

I said, "The password's—" Again the line went dead.

I called again. It rang once. Dion picked up. "Give it to me quick!"

I didn't even get out a syllable when the line went dead for a third time. The next time I called, only a few minutes later, a female-sounding android informed me that the phone number had now been disconnected.

Dion tried calling me collect, but he would drop off the line after only a second or two. As a test, I called my friend Genevieve in Lakewood and had her call me collect. Of course, her call came through fine. Again I tried calling Dion. Again we were cut off after a second or two. Any doubt I had about the reality of any of this nonsense flew out the window. Listening to the dial tone beeping into my ear, I tried to figure out some way I could get the password to him without being interrupted again.

Then I remembered something unique about collect calls. I hung up and dialed the operator. "May I help you?" the operator said.

"Yes," I said, "I'd like to place a collect call to Dion Fuller at 858-273-5192."

She dialed the number. It rang once. I heard Dion say, "Hello?"

The operator said, "Mr. Fuller, I have a collect call for you. Do you accept the charges?"

Before Dion could say anything I yelled out real fast, "*TheMexicanguywiththethirteeninchdick!*"

We had once known a murderer, drug addict, and member of the Mexican Mafia who complained bitterly to us back in 1998 about the enormous burden of having been born with a thirteen-inch penis—thirteen inches when *limp*. The man even felt the need to show it to us. This was not something one was likely to forget. The man's name was Chino.

I heard Dion laugh for a half a second, which was then cut off by the line going dead once more.

I hung up the phone and collapsed onto the bed. Mission accomplished.

I sat there staring at the ceiling and stewing. Those sons of bitches are really doing this, I thought. This wasn't some kind of weird meth-induced psycho-drama. It was really happening. There was no other explanation.

I picked up the phone and called Wanda, skeptic of skeptics. I'd known Wanda since high school. So had Dion. In fact, Dion dated Wanda when he was sixteen and she was about thirteen. Now she was a recovering heroin addict (and a recovering mental patient to boot) who had two kids and worked as an emergency room nurse. I had been keeping her updated on the unfolding saga ever since it began, but her opinion was that Dion was just off his nut. So I called her up to tell her what had just happened.

"How do you explain *that*?" I said.

"How do you know Dion wasn't just hanging up on you?"

I rolled my eyes. "Jesus, Wanda. First, it wasn't the sound of someone hanging up on me. It was the sound of the line actually going dead. And besides, you know how Dion is with money. You think he could wait even *two* extra minutes if he knew there was seven hundred fucking dollars waiting for him at a Western Union just across the street?"

"Well… maybe you're right. What's the name of this woman who's supposedly harassing him?"

"Do you even pay attention when we talk? I told you before. *She's Special Agent Lita Johnston.* I gave you her name, address, and phone number a couple of weeks ago." Dion had read the information to me off her business card. I later read the information to Wanda just to prove the woman existed. She thought Dion was making *her* up too!

"I'll call you back," Wanda abruptly said.

"Okay," I said. I figured she had to go deal with her kids, so I just went back to writing whatever I was writing at that moment and waited for Dion to call back. I hoped he would be able to throw everything into his van and get out of town before they prevented him from leaving.

A few minutes later my phone rang. It was Wanda.

"I just talked to Lita Johnston," she said.

"You did *what*?"

"I called that number. You know, that woman really exists. She works for the NCIS."

"I *know* she really exists! I never doubted it! *You* were the one who was doubting it!"

"Yeah, I know. That's why I called her. So I asked her if it was okay if Dion left town."

"You did *what*?"

"She said he wasn't under suspicion for anything. Then I asked her if anyone was watching him. She said no."

I couldn't believe what I was hearing. "You're joking with me." I could hear her two kids crying in the background.

"No. I really spoke to her. She seemed nice."

"I can't believe you did this. Why did you do that?"

"I just wanted to help out. I explained to her that Dion didn't know anything about those night vision goggles. She wanted to know how I knew about that."

"What'd you say?"

"I told her *you* told me. For some reason she couldn't wrap her mind around the concept that someone could be living in Dion's house with stolen items without Dion knowing it. I tried to explain to her what the life of a junkie is like, but I don't think she was buying it. I tried to invite her to one of my NA meetings so she could better understand the situation, but she didn't seem that interested."

"He's dead. You've just killed Dion. Now they know he's planning to leave."

"Well, if they're listening to his phone calls they already know that, right?"

"I... well, yeah, I guess so."

"No harm, no foul. Don't worry about it. If Dion would just stop doing drugs, then he wouldn't get mixed up in situations like this."

"Wanda, I understand that none of this would've happened if he hadn't been doing drugs, but that doesn't give the U.S. government the right to randomly persecute innocent civilians just because they think it's fun."

"It doesn't seem random to me. Dion was asking for it. I'm glad the U.S. government has these jokers under surveillance. Aren't they supposed to protect us from people like Dion?"

"Who's Dion harming except himself? What you're advocating is completely at odds with the U.S. Constitution."

"If it works..."

"Listen to what you're saying. If you're accepting that everything I've told you is true, and has really happened, then that means the U.S. government has access to weapons and technology that's about fifteen to twenty years beyond the capability of the private sector. They could be using that technology for so many useful things. They really *could* be using it to prevent violent crimes, etc., but they're not. They're using it to spy on *Dion*. On fuckin' *Dion*! *That's* the best they could do with it? If they've got the technology Dion's described, there shouldn't be any more crime throughout the entire United States. Is that what's happening? No! Crime's going *up* and in some office in the NCIS in San Diego is a complete file on Dion's masturbation habits!"

"Listen, I don't know the answers to these questions. I've got two kids to deal with and I'm kickin' my boyfriend out of the house. I was just tryin' to help out."

"Thanks, Wanda. You helped out."

"I'll call you back to see what happens. Tell Dion I said hi." This last sentence was uttered without a trace of irony or sarcasm. She rang off, then I sat there staring at the phone for a while. I tried calling Dion, but his phone just rang and rang. I dug out the spiral notebook in which I'd jotted down Lita Johnston's contact information. I stared at it for a few seconds, my Imp of the Perverse tugging and clawing at my curiosity. Then I grabbed an old Walkman cassette recorder out of my closet, set it up by the phone, hit the speaker button, and dialed Lita's phone number.

What follows is a complete transcript of that phone call:

LJ: *[upon answering the phone]* Lita Johnston.

RG: Hello, Ms. Johnston. May I speak to you for a moment?

LJ: *[slight pause]* Yeah, who's this?

RG: This is Robert Guffey. I'm a friend of Dion Fuller.

LJ: Yes?

RG: Can I just ask you a couple of questions, please?

LJ: Okay.

RG: I was just wondering, is he currently under investigation?

LJ: Well, this is not something I generally talk to the public about. I've had several calls already from other people regarding this matter. I don't really want to discuss it any further.

RG: I understand that. But he's under the impression that he's being followed or something.

LJ: Well, sir, I already spoke to Mr. Fuller about this several, *several* times already. So if he continues to think that, then there's nothing I can do about that, sir.

RG: Is there a warrant out for his arrest?

LJ: Sir, I really can't discuss this any further. It's really a matter between, uh, Mr. Fuller...

[Pause]

RG: Between Mr. Fuller...?

LJ: Mm-hm.

RG: And who?

LJ: And *myself.*

RG: Is it legal for him to leave San Diego?

LJ: *[Pause]* Well, sir, he's not under arrest now, so he can do as he wants, but I wouldn't advise him to. But, you know, that's a decision he's going to make.

RG: But why wouldn't you advise him to if it's legal?

LJ: *[Long pause]* Well, sir, as I said, he's not under arrest at this *moment*, so if he chooses to leave I can't stop him from doing that.

RG: I just don't want him to do anything illegal.

LJ: Mm-hm.

RG: So if he was planning on leaving, and it was illegal, I wouldn't want him to do that.

LJ: Right. I understand. Well... at this point it's not *il*legal, but it would just complicate matters. Potentially.

RG: I see. Okay, well, I just wanted to ask you those few questions.

LJ: Okay. What was your name again, sir?

RG: My name is Robert Guffey.

LJ: Robert Guffey?

RG: Guffey.

LJ: Oh, okay. You live here in San Diego?

RG: No, I don't live in San Diego. You see, I don't really understand the situation because I don't live in San Diego.

LJ: You're just a friend of his?

RG: Yeah.

LJ: Okay. Did he ask for you to call me?

RG: No.

LJ: Okay. I was just curious how you got my number.

RG: He gave me the number.

LJ: Okay.

RG: But he didn't ask me to call you.

LJ: All right, well… that's about it, sir!

RG: Okay, thank you very much for your time.

LJ: Thank you. Bye-bye!

I immediately called Wanda and played the conversation back to her. "You hear this? I mean, if that's not a bunch of cover-your-ass-speak, I don't know what is."

"Well," Wanda said, "maybe they are really watching him, but I doubt they're sending invisible midgets after him."

"If agents of the U.S. government are clearly following you around and harassing you 24/7, why would you feel the need to make up the part about *invisible midgets*? Wouldn't the whole 'harassed by the U.S. government' part be dramatic enough?"

"Maybe he really believes it."

"I've had a billion conversations with him since last July. I *know* he believes it!"

"Then you agree with me, he believes it."

"He believes it because it's probably *happening*."

"Whatever, I have to go." She was right. I could hear the kids crying in the background again.

I hung up, then sat there staring at the phone, dreading having to call Dion back and tell him that Wanda opened her big yap and told Lita he was intending to bail. On top of that, I would then have to tell him I exacerbated the problem by calling Lita myself and essentially telling her the same thing. Why the fuck did I do that?

I picked up the phone and dialed Dion's number.

Dion seemed out of breath when he picked up the phone. "I've almost got all the shit I want to take with me in the back of the van," he said. "Now I just have to go over to the DMV and get the registration in my name. I don't know how long that's going to take. I hope I can get out of here before sundown. I made the mistake of telling my dad I was leaving. I stupidly gave him Lita's phone number a few months ago, hoping he could use his ex-cop powers to do something about all this, so of course he picks today of all days to call Lita and ask her what the fuck is going on. The one day I didn't want anyone drawing Lita's attention to me is the day my dad picks to finally get up off his ass and do something about all this fuckin' nonsense!"

"And what happened?"

"She just blew him off. Said I was crazy."

"That's funny," I said, "you know what Wanda just did?"

So I told him.

Dion blew up.

"What the *fuck*? Is that bitch trying to get me killed? Is everyone in the world against me?"

I decided to leave out the whole part about me calling Lita back. "That Wanda," I said, "what a *bitch*."

"Damn straight," Dion said. "Now I've got to haul ass even quicker to get out of here before they send out the invisible hellhounds. You should never have given Wanda that number, man. Jesus fucking Christ. I think I'm gonna have a heart attack!"

"Take deep breaths. Through your nose."

He did exactly that. "Thanks, man. I think I'm okay now."

"Hey, where're you headed after you've gotten everything in the van?" I hoped he didn't say, "Your place."

"I'm headin' west," he said with a smile in his voice.

Then he hung up.

West from his apartment was the Pacific Ocean.

11.

Dion managed to get out of town just before sunset. He called me back one more time to tell me that he went over to his landlord's apartment to tell him he was leaving at long last.

The landlord looked flustered and ashamed. He breathed a sigh of relief and said, "Thank you, man, thank you."

"Why're you thanking me?"

The landlord looked suspicious as his eyes darted from side to side, as if afraid that someone was listening. "That just makes everything a whole lot easier for me. You know."

Dion wanted to ask him further questions, but knew that would be useless anyway, and instead just tossed him the keys to the apartment and got back to packing. On the way out of town he stopped by Jessica's apartment. He knew she was now living with some new paramour on the other side of town. He desperately wanted to give her his cat, the cat they had bought together. Over the last few months it had gotten sicklier and sicklier, losing a lot of weight, just like Dion himself. Unfortunately, Jessica wasn't home. He considered picking the lock on her window and just leaving the cat in her living room, but he couldn't bring himself to part with it, so he drove out of town with the cat still in the back of the van. I'm not even sure he had food for it.

He left the city limits of San Diego with five cars on his trail. I wondered if they would even let him leave. But they did. And as he sped farther and farther away from San Diego, those particular five cars turned around and drove away.

From that point on he would call me from his cell phone every other day to give me updates on his progress. He said he was pretty sure that no vehicles had followed him since those initial five. However, he insisted he kept seeing some kind of circular device in the sky that appeared to be trailing him.

"What do you mean?" I said. "Like a flying saucer?"

He said it was kind of like a disc—but relatively small. If not for the sunlight glinting off its metal surface, he never would have noticed it. Sometimes it was there, sometimes it wasn't.

In the early 1980s the Space and Naval Warfare Systems Center in San Diego developed vertical-take-off-and-landing air vehicles known as Airborne Remotely Operated Devices (AROD). In the 1990s these vehicles evolved into robots called Multipurpose Security and Surveillance Mission Platforms. According to the corporation's own website, such robots were "designed to provide a rapidly deployable, extended-range surveillance capability for a variety of operations and missions" including "tactical security" as well as "support to counterdrug and border patrol operations."[10] Since residents of California are well aware of the fact that the border patrol can't seem to keep people from slipping over the border into San Diego, perhaps the real purpose of the MSSMPs is to keep certain people *in*.

Note: The San Diego headquarters of the Space and Naval Warfare Systems Command is located only six miles from Dion's Pacific Beach apartment, a mere ten minute drive.

12.

During his second day on the road, in the middle of the desert, Dion pulled off at some desolate roadside rest stop to rearrange some boxes in his van. At the soonest opportunity, the cat jumped out of the van and took off into the desert. Dion tried to follow, but the creature was too fast for him. The animal had been a domestic cat its entire life. Dion called me, almost in tears, and said, "Don't you think someone probably found it and took care of it?"

---

10. http://www.spawar.navy.mil/robots/air/amgss/mssmp.html

In the middle of the desert? It was probably eaten within the hour. "Uh, sure," I said. "Why didn't you just leave it at Jessica's place like you said you were going to do."

"What if Jessica was out of town? I didn't want it to be alone."

I think what he meant was that *he* didn't want to be alone.

Dion noticed that whenever he would drive within the general vicinity of a military base, unmarked vehicles would begin tailing him again. It was the same pattern as in San Diego. It was impossible *not* to notice them. He wasn't sure they were even trying to be stealthy about it.

As he drove farther and farther away from the base, the vehicles on his ass faded away. At this point he called and asked me to find a map of all the military bases in the country so he could avoid them. I found one, and quickly discovered that it was impossible to travel across the country without finding yourself in the general vicinity of a military base at some point or another.

Eventually, he found himself in Alpine, Texas. He hadn't called me in days and during that time hadn't experienced any harassment at all. Three minutes after he hung up with me, five vehicles appeared out of nowhere and began tailing him again. From there he went to Austin, where he somehow met Hunt Sales, punk rock legend and son of Soupy Sales. Dion stood in a parking lot from midnight to 2:00 AM and told Hunt Sales the entire story you've just read. Hunt told him to go to NA.

From Austin Dion drove to Arkansas, where he ended up in indentured servitude to a sadomasochistic gay priest, then asked me to call his ex-girlfriend Melanie, a stripper and former punk rock musician now living in New Orleans, to let her know he was going to try and stop by and see her. Unfortunately, he didn't have her phone number, so I had to call every single Melanie McGill in Louisiana until I reached the right one. To my surprise, I hit it on the second try. I discovered that Melanie now lived in Harahan, Louisiana, on Crislaur Street with some budding rock star named King Louie. The second I got ahold of her I launched into the story, invisible midgets and all. Melanie's reaction was to laugh uproariously, but she didn't hang up. She wanted to hear the whole tale. Her opinion seemed to be that Dion was batshit crazy. I didn't really say anything to curb this opinion. I would say things like, "Invisible midgets, what a card, that Dion!" When I told her Dion was on his way to Louisiana the phone went silent for a second, then she begged me to do anything I could to dissuade him from getting anywhere near Louisiana or her house; nonetheless, she did want me to call back and keep her updated on whatever happened to Dion because she said it was the most entertaining TV show she'd seen in a long time.

The next time Dion called I didn't have the heart to tell him that his ex-girlfriend thought he was nuts and that she never wanted to see him again. Thankfully, I didn't have to. Before he could get anywhere near Melanie's house, he was arrested by the New Orleans police for being drunk in public (to Melanie's great relief).

Unlike with Melanie, this entire imbroglio seemed to have a positive effect on Dion's dad. For the first time in over a decade, his dad seemed sympathetic to Dion's problems, so much so that he asked me to send his son money via Western Union whenever Dion asked for it and assured me I would be reimbursed with a check as soon as possible. And he was good to his word. I always received a check for the proper amount within a few days of having wired Dion the money. Why Dion's father couldn't send him the money himself confused me.

The same exact day Dion got himself arrested, Joan Sienkiewicz called to let me know that she had contacted a lawyer in Minneapolis (coincidentally, the very place Dion wanted to reach) who was willing to listen to Dion's story. The lawyer's name was James Anderson; he'd had experience representing survivors of ritual abuse/torture. I thanked her for all her efforts, and didn't mention the fact that the guy she was going out of her way to help was now collecting dust in a New Orleans jail.

13.

Meanwhile, I was trying every desperate move I could think of to pick up classes at either CSU Long Beach or on another campus, so I sent the following email (titled "The Curious Affair of the Invisible Midgets") to Dr. Eileen Klink, the chair of the CSULB English Department:

Dear Dr. Klink:

A close friend, who until recently was living in San Diego, is currently being pursued across the country by a covert team of invisible midgets and/or very slender acrobats funded by an unknown government agency. This is an incredibly long and improbable story, too complicated to go into in an email as short as this, so I'll give it to you in broad strokes.

I'm trying, as much as possible, to prevent my friend from being tortured by highly sophisticated electromagnetic devices known in intelligence parlance as "nonlethal weaponry" (turn to page M1 of the 1-4-04 edition of the L.A. Times *and read the article entitled "Pulling Punches" by William M. Arkin for more information). Such technology includes mind-altering, damaging weapons called "acoustic bullets" (see <http://abc-news.go.com/sections/wnt/DailyNews/sonic_bullet020716.html>). Since this unnamed agency is currently employing flying probes called ARODS (see <http://www.spawar.navy.mil/robots/>) for the purpose of keeping very close tabs on my friend, it's quite difficult for him to attain even a moment's* *respite without the chilling shadow of the law always looming over him. Apparently, such peace can only be attained with a fair amount of money in order to hire paranormal entities known as "lawyers" and "private detectives" to prevent these midgets and/or very slender acrobats from having their way with my friend.*

Since I'm gearing up to wage a difficult legal battle with these people, a battle I'm sure you would support morally at the very least, any extra cash would help. Hence, this email, inquiring into the present state of affairs concerning English 100 teaching positions available in the spring semester. In the event that anything at all is available, I would be very appreciative if you could keep me in mind for a position. Otherwise, my friend might end up being star-chambered and gaslighted by invisible midgets and/or very slender acrobats and we don't want that to happen. (For evidence concerning the advent of invisibility, i.e., "light-bending," technology please click on this link: <http:// www.eetimes.com/at/news/OEG19990120S0003>).

Remember, we're all in this fight together.

Sincerely,
Bro. Robert William Guffey, 3º

For some reason the chair did not respond to this message. My fears were no longer unfounded. The invisible rapscallions were clearly preventing me from working. Dion's problem was now my problem and always had been. I had to continue doing whatever I could to help him fight the oppressor.

The second the cops released Dion three days later, he hightailed it back over to the parking lot where he'd left the van. He was shocked and relieved to see that it was still there. He must have made quite an impression on the New Orleans police, as they told him to haul his ass out of town the moment they let him go. So, from Louisiana he went to Minnesota to drop off a package for his teenage son. From there he decided to visit an old friend from high school, a girl named Peru who now had a baby and lived with her boyfriend, some shady dude named Jawbone. Peru and Jawbone allowed Dion to stay at their house for a night.

Around this time, the service on Dion's cell phone was canceled, so he had to start calling me from pay phones. Some of these calls I recorded.

CONVERSATIONS WITH DION, PART 1
(3-10-04)

RG: ROBERT GUFFEY
DF: DION FULLER

DF: When you hung up the phone with me, my phone starts going… you know, it sounds like it's giving change back or something?

RG: Uh-huh.

DF: And I pick it up, nothing. Just, like, totally flat dead. And I'm like, oh, that's bizarre. Then I pick up the other phone that I'd given you the phone number to… same thing starts happening. But I hear the machine go, "Hold on. One second. Please wait one second. Please wait one second. Please wait one second. Serial number: 8-1-8-6-4…" Then it goes, "Thank you! 9-4…" And then it's just starts reeling off all these numbers, and then it does it again.[11] What're you doing?

11. In Chapter Nine of his classic 1975 work of paranormal investigative journalism, *The Mothman Prophecies*, John A. Keel mentions people whose phone conversations are suddenly interrupted by voices reeling off random numbers. In March of 1961, for example, two women in Prospect, Oregon, had their phone conversation interrupted by a man's voice reciting "the numbers forty and twenty-five over and over." Keel elaborates:

RG: I'm recording this. I'd like you to take it from the top. Can you say the first thing that happened in Minneapolis?

DF: Okay, man. I've got twenty-three minutes on this.

RG: Okay, just talk about Jawbone... what they said to Jawbone...

DF: Oh, uh, well... what the cops said to Jawbone... was, uh... they asked if I was living with him. He's like, "No." They asked him about the stuff I had moved in and out of the house. They told him that I'd been talking to people, and I'd been talking to people about "doing something." They questioned him for forty-five minutes. They asked him to go down to the station. He said he couldn't do that, he had to go to work. And, you know, he's forty-five minutes late. And just various questions about me and implying that possibly I'd be some sort of terrorist.

RG: They actually used the words—

DF: No.

RG: But they said, quote, "He's been talking about doing something." Unquote. That's it.

"Voices counting off meaningless numbers also cut in on TV reception in UFO flap areas. Usually people who experience this sort of thing dismiss it as police calls or the work of some Ham radio operator. They don't realize that TV sound is broadcast on FM channels reserved for that purpose and there is little chance that a shortwave or CB (Citizen's Band) transmission could interfere.

"But the phenomenon is not always restricted to electrical apparatus. After I published a couple of pieces about it I received dozens of letters from people throughout the country recounting their own experiences. To my surprise, most of these people had heard the voices late at night, usually waking them up with a sharp command. For example, a man in the Southwest claimed he had been jarred awake on several different nights by the sound of a deep male voice ordering, 'Wake up, number 491!' A woman in Ohio heard the voice while driving, '873... You are 873.' And another woman in Kansas wrote, 'Please tell me who these people are that keep reading numbers to me. They sound as if they are standing right next to me but there is no one there.'

"Do we all have a number tattooed in our brains? Hardly. There are three billion people so some of them should be numbered 2,834,689,357. But all of the numbers that have come to my attention contain only two or three digits."

DF: Yeah. "He's been talking to people about *doing* something." Period.

RG: And where did they talk to Jawbone?

DF: Well, he walks to work every night. They picked him up on the way to work. Which cracked me up because Peru got all pissy with me, and I'm like, "Look, they obviously know you guys." She's a weed smoker. It's obvious that, you know, they've been under surveillance. And she mentioned that, because of some rapist in the neighborhood, they've picked up Jawbone a few times and shit. But this was all questions about me, my comings and goings. And the fact that I was talking to people. I haven't talked to anybody in three days, except for the lady at the food stamp office. Other than that I've been living in the van, reading, and going to a shelter to eat.

RG: And there's no way the woman at the food stamp place called or—

DF: Not at all. No, this happened the night before I went to the food stamp office.

RG: Okay.

DF: No, she was totally sympathetic, and understands that they waste their fuckin' money on a lot of stupid shit.

RG: *[Laughs.]*

DF: *[Pause]* Uh... would you be... would you be able to send some money? And, uh, I talked to my dad. Actually, he's calling you tomorrow to get my license plate number off the van.

RG: I just sent it to him in an email.

DF: Oh. Did he already reply?

RG: No, not yet. I emailed him a few hours ago.

DF: Yeah. He'll probably check that, but if not he's gonna give you a call because of all this bullshit.

RG: Yeah, yeah.

DF: It sucks. I wish you were recording that, like, *way* earlier. I meant to say something when I first got on the phone as well, because this needs to... I mean, I really don't think I'm going to be living much longer. There's a bunch of nervous-ass cops with guns all over me and my car. And just lying and... funny, after all this bullshit with Peru happened I went back to this gas station where there was this cute little punk rock chick workin'. And I'm frazzled out of my mind, just driven to the point of complete anxiety, and I'm, like, "Hey, listen to this." And I go into this whole story with her, and tellin' her all this crap. And I'm like, "Look, I'm the target of continual government, like, *harassment*, and blah blah blah." And, you know, I was in there earlier and asked to use their phone because my phone card wasn't working properly and had a normal old everyday conversation with her, so when I was talking to her about this seemingly crazy shit... she believed me. And had no problems with what I was saying. Then I moved my van across the street because the guy delivering the gasoline to the gas station needed me to move. And so I moved it across the street. This girl, after hearing my story, gave me five bucks for gas. I went to go get my van from across the street to pull back into the gas station, and in the short amount of time that that took, two cops passed... well, one guy in a red car passed by and put his hand to his mouth and obviously said, "He's moving." And as I went to turn into the gas station a cop came up right behind me and said I was being questioned because I didn't use my left turn signal. And as I sat there, in less than thirty seconds, four other squad cars pulled up, one equipped with a drug dog. And they asked if they could search the car. I said, "Nope. I live in there and I don't want you going through my personal belongings." They go, "We thought you'd say that." They asked if I was in a militia. I asked why, and he's like, "Because of your shoes." *[Laughs]* I said no, I drove for an ambulance company in Inglewood. Another cop comes up, who I found out later was the sergeant, and goes, "You lived in Inglewood?" I go, "No. I drove for an ambulance company in Inglewood." The other cop who pulled me over asked if I still *did* that. I go, "No. I'm in Minnesota right now."

RG: *[Laughs]*

DF: They're dumbasses, man. Then he points to the temporary registration tag on me and he goes, "I thought you said you got here two days ago." I go, "I did." He's like, "This says January 6." I go, "Well, that's when I left." And he goes, "Well, now your story's changing. I thought you said you got here two days ago." I'm like, "No, that's when I *left*. I got here two days ago. That's correct." And then when I said I'd spent time in the New Orleans jail, they didn't want to hear any of that. Eventually, when I was sitting in the cop car, with my lengthy criminal history, the only thing that came up were the drug charges that were dropped in San Diego.

RG: *[Laughs]*

DF: There was a printout of all the charges I was charged with in San Diego. Not my burglary. Not the possession for sales. None of, like, the serious crime came up, just the drug charges.

RG: It didn't mention the fact that you were in jail in Maryland back in '98 or—?

DF: None—*none* of that. Just, bam, drug charges and San Diego stuff. I was arrested in New Orleans just a fucking couple of weeks ago. None of that, just the San Diego shit. When they asked, "Have you ever been arrested for a drug charge?" I'm like, fuck no, because I *haven't*... unless you count that pharmacy worth of shit I stole.
[Both laugh]

DF: No, that didn't come up, you know. *[Laughs]* And right off the bat one cop comes up and starts saying shit about crystal meth. And I'm like, "What the *fuck're* you talkin' about?" You know? And they're like, "You're acting pretty fidgety."

RG: *[Laughs]*

DF: "My partner over there says you're acting pretty fidgety." And I'm like, "Fuck, I just had a bunch of cups of coffee." And they're like, "Maybe, *duh-duh-duh.*" They knew *all* this crap about me and shit, man. But… can you *please* send money?

RG: When?

DF: Tomorrow. Like as soon as possible. And the code is the name of the guy who had been good friends with Philip K. Dick who they executed in the desert.[12]

RG: Okay.

DF: Just the first and last name.

RG: Okay.

DF: J.P.

RG: Right, okay.

DF: Yeah, I got his book today.

RG: Which one?

DF: *The Other Side.*

RG: That's a good book.

DF: It's in excellent, like, pristine condition. I saw that and L. Ron Hubbard's… something *Dreams.* Like, one of his science fiction books. That was in pretty excellent condition too.

RG: There might be some hypnotic triggering cues in there. *[Laughs]*

DF: Yeah, that's what I was figuring! But actually I just left that. I just wanted the Pike book. I figured that was more relevant to the situation.

RG: You know, Pike was in the Navy.

---

12. Bishop James Pike.

DF: Navy Intelligence. *[Laughs]*

RG: It would be easier on me if I could... I mean, I have almost *nothing* left in the account.

DF: Oh, do you?

RG: It would be great if your dad's check would come first, *then* I could send it to you.

DF: I have *nothing*. I can't even get out of here. If anything, just for me to be able to get away from this shit, you know? But it should be there. When my dad went to the mailbox and sent me the stuff that's coming to Peru's, he sent it out on Wednesday. The thing didn't come to Peru's on Saturday, so I'm sure your check's gonna be at your house tomorrow, you know.

RG: But what're you gonna do? It's, like, it keeps *happening*...

DF: Fuckin', I have *no* idea. I can't even get work. I got the majority of stuff to, like, keep me afloat with food and stuff. I just need to be able to take a shower and put, like, about a hundred bucks into the car for a tune-up. And a tank of gas. Then I can get to, like, you know...

RG: Yeah.

DF: Yeah, I don't fucking know, man. I mean... but, I mean, I need more than that because I don't want to have to call... like, I want to talk to you, but I can't call and my dad's like, "Look, don't tell me where you're at."

RG: Yeah, yeah, yeah.

DF: And it's, like, you know, they're tracking all this bullshit through the phone.

RG: Yeah.

DF: *The phone. [Pause]* And stuff. You know. And I know I gave you and my dad Peru's address. You didn't send anything, did you?

RG: No.

DF: Good. Because that's just... man, she just flipped, and, oh my God, this just sucks.

RG: I mean, Joan said that a lawyer could help you, but the problem is you're mobile and *[laughing]* you have no money...

DF: Well, I'm headed back that away. You know, I'm headed to Haiti. I'm gonna go figure out that whole situation for 'em. Settle it for 'em.

RG: Uh-huh.

DF: I figure I'd be pretty good in that kind of situation.

RG: *[Laughs]*

DF: *[Laughs]* Know what I mean? So, like, yeah, when I'm able... and that's what I want to do... but, uh, that address you gave me, I'm gonna get in touch with that person...

RG: You should. Yeah, I would advise that.

DF: Oh, yeah. Could you look up asylum stuff on the Internet for me?

RG: Asylum?

DF: Yeah, in Canada.

RG: Oh, I see. Okay.

DF: No, not like the one Wanda was in.

RG: I'm sure they could provide that for you. *[Laughs]*

DF: *[Pause]* Well... they do. They used to do that sort of shit.

RG: No, I meant the *other* kind of asylum.

DF: Oh, yeah. *[Laughs]* Okay. No, I want to look into that. Because this is just insanity. It was worse than San Diego. I never had fuckin' eight million cops on my ass like that. And they were just like... just such off-the-wall shit was coming out of their mouths. Now I'm at this truck stop. Oh, you should've seen the way—there had to be, like, forty cars that were behind me. I swear, Robby, it's just so fucking insane. I would, like, speed up and go toward an exit and stop right before it so all the cars on the freeway would have to pass me. And then, like, half of them would get off. The other half would keep going. And then I would go and do it again. The half that had gotten off had come back behind me *again*. And we were just going over and over it again until, like, I got… you know, the majority of them had been, like, displaced onto all these different exits. I pull over onto this truck stop and I say to this guy, "Look, I'm a U.S. combat veteran from the first Gulf War. A close friend of mine from the first war stole all this shit from the military, and basically they think I have something to do with it. I just got escorted out of Minneapolis and now I'm just gonna sleep in your truck stop parking lot. Is that okay?" Guy's like, "Yeah, no problem, man." I go, "If anybody comes into here and says anything about me *please* let me know." Dude, you know, they couldn't say that, like, "Hey, are you an anarchist? You this, you that?" Because, dude, I'm wearing work pants, boots, and I have a crew cut. I look like I'm fuckin' full-on military.

RG: Why did the guy say, "We thought you were in a militia because of your *shoes*?"

DF: Because of my shoes. Because they're Special Forces shoes.

RG: I see. So, he was just grabbin' for straws.

DF: Oh, yeah! You know, he expected some, like, litany of Nazi shit to come out of my mouth because *that's what they've been told.*

RG: Right, right, right.

DF: And shit. They go, "Oh, okay, we're going to give you the van. We don't want you walking around Richfield all night, blah blah blah." I go, "Why did you do this, and how did you get this information?" He goes, "What do you mean?" I go, "Well, a friend of mine told me he was talked to by—" And he goes, "*What's his name?*" And I go, "Well, a *friend* of mine—" And he goes, "Well, that's not true, I just met you this evening." And he goes, "What's his name?" And I go, "Well, never mind." I just, like… you know, went about my business. Oh, it's just insane. Just ridiculous. The amount of… when I went into that, uh… *[Pause]* So

how much? And I'm in Minnesota. And, uh… this area…
'cause this phone card's gonna cut off. I have to get another
phone card, but…

RG: Is it possible for your dad, just this one time, to go—?

DF: He *can't*… well, there's, like, no way… he has no problems
sending *you* the money. He can't because he has to do that
on a credit card.

RG: Why can't he do it by cash?

DF: I—dude, dude, it's my *dad*. I don't *know*. You know,
I mean, I *told* him. I guess whenever he leaves he's with
Linda *[Dion's stepmother]* or something. And it's easy for
him to put the shit in an envelope.

RG: I see.

DF: Is he sending you cash?

RG: No, no. He's been sending me checks.

DF: Dude, I don't get it, you know. And it's, like, each time
I'm in this situation it's, like… I'm in it because it's, like,
fucking *dire* crazy straits, you know?

RG: Of course.

DF: Yeah, and there's stuff that's coming from my dad's that's
taking mysteriously longer than it should, stuff that was
going to Peru's house.

RG: Uh-huh.

DF: It was my license, a copy of my birth certificate, all this
really important stuff that I'm unable to get because they
made me *leave*. But through the food stamp office—which,
like, pissed them off to no end—I've got the food stamp
card, I've got various forms of ID to where I could go get
work.

RG: Yeah, that would be good.

DF: Yeah, and that's like, you know… I was gonna do that
tomorrow morning in Minneapolis. I was, you know,
escorted the fuck out of there. I haven't taken a shower in
a week. I'm like… I feel like *crap*. I'm like, you know… it's

crazy. I'm not on drugs. I haven't dranken anything. It's just, like, I'm fuckin' worn *out*, man. The other night… like, after I found out the stuff from Peru, I start the car. I'm on a totally isolated, deserted street and it's snowing. I hear people, like, walking outside… it's funny, because all the shit about midgets—

RG: *[Laughs]* Yeah.

DF: —I've noticed, I've seen more midgets than *anything*. I think they think I have some, like, fear of midgets or something.

RG: Not when they're visible. *[Laughs]*

DF: Exactly! That's what I was thinkin'. I don't care when I can see 'em. *[Segues into an Edward G. Robinson impersonation.]* "I'm big, see!"

RG: *[Laughs]*

DF: It's the ones I can't see that I'm scared of. Uh, but, yeah, there's like, this person, like, as I'm coming down the street, was hurriedly, like… like, rapidly trying to pull this midget female out of her house and she was obviously *wasted* drunk. And I stopped. And the person was trippin' out that I stopped. And I'm, like, wavin' 'em by. 'Cause I thought it was an old lady at first. But then I realized it was a *midget*. Oh man, I got these glasses out of a thing that people donate their eyeglasses to? And, man, this is better than—I can see a fuckin' hemorrhoid on a gnat's ass with these glasses, man. It's an *amazing* difference. I can see *everything*. It's cool, and they're, like… they're, uh… L.A. Gear, and they're, like, nice, nice glasses.

RG: Oh, good. They're not serial killer glasses?

DF: Oh, no. No, no, no. They're nice. No, I look like, fuckin'… not at *all*. No, I mean, they're real—they're kinda like your glasses, I guess. Smaller frames—

RG: Yeah.

DF: —and just really nice. I look edjamacated—

RG: *[Laughs]*

DF: —and shit. But, uh… oh! 4:30 in the morning, in the snow, freezing outside… I wake up randomly at a weird time, so I hop in the front of the car and I just start the car. And I notice behind me—because I'm watching behind me— two sets of headlights come on. Then a *third* car comes around the corner, and there's a space behind me, and tries to pull in, but at a weird angle. And it's this woman who looks kind of confused, like "What do I do?" type look?

RG: Uh-huh.

DF: So I roll down my window *[laughs]*, and I wave her by. Like, "Hey, go by, I'm not leaving." And she gets this look on her face like, "What the fuck? What's this guy doing?" *[Laughs]* And she goes and pulls away. And if it was some- body who was looking to park right there, just randomly at 4:30 in the morning, they would've came and parked further up the block, correct?

RG: Yeah.

DF: Or *come back around* to see if I had left yet, correct?

RG: Yeah.

DF: Nope. All three cars, *bam bam bam*, leave, like, right by me as I'm, like, flippin' 'em off. *[Laughs]* Not, like, out the window, but just, like, inside of the window—

RG: Yeah.

DF: —so they have to look over to see it. *[Laughs]* And not one person stopped… nothin'… and shit. 4:30 A.M. And, uh… the cop goes, "Man, why did you have to come out here tonight? At this time of year? It's fuckin' *cold!*"

RG: *[Laughs]*

DF: And he smelled like booze, too. Which… I was gonna say somethin', but then I thought I'd just get my ass beat.

RG: Well, you know, you have to get through that below-thirty weather somehow.

DF: Yeah, exactly. All smellin' like fucking alcohol. Uh, it sucks. I need to sit down with you and go from the beginning of this conversation, like *everything*… because you could fill in, like, the blanks when I start—

RG: Uh-huh.

DF: —the story. Because I really need to get this documented on tape.

RG: Yeah, well, maybe we can do it through the *phone*. *[Laughs]*

DF: Yeah. *[Laughs]* Oh, yeah. No, we're gonna do it down on the beach in San Diego.

RG: *[Laughs]* Along with all the midgets there?

DF: Yeah. Exactly.

RG: Have you seen anything more like that recently?

DF: Oh, it's still—oh, yeah. The predator drones are, like, my best buddies out here.

RG: But you haven't seen anything like *that*. Like the invisible stuff? *[Pause]* Or have you?

DF: I get the *feeling* every once in a while.

RG: Oh, okay. But it's not like it was so obvious before where you could actually *see* it?

DF: No, not so much, no.

RG: Yeah, okay.

DF: Yeah, I've been trippin' on that kinda. But… oh, what I *have* noticed is that they have the same kinda thing for vehicles where, like, you look at the car and it looks like there's nobody in there. And it, like, actually looks like there's depth, and you could reach in there and there'd be nobody there. And I've seen… they've used them all over the place in San Diego, but in fact if you get up *really* close you could see the silhouettes of people sitting in the cars.

RG: Oh, you mean, from the way the windows… it looks like there's nobody inside?

DF: Yeah, but it's a hologrammed image where it looks like… you know, like the Fakespace stuff. I finally went through all my email, by the way—how much could you send me by tomorrow? *[Laughs]*

RG: *[Laughs]* Well, I couldn't send more than 150 without closing my account.

DF: Okay. That's… that's… that'll work.

RG: Okay.

DF: And my dad will… I'll get on his case tomorrow.

RG: Okay. I wish your dad would not be so fuckin' pussywhipped.

DF: Tell me about it, man. And it's like… you know, I think he kinda digs on the spy-weird thing or something?

RG: *[Laughs]* Then why does he make *me* do all the spy stuff?

DF: I know. It's fuckin' stupid. No, I know he needs to keep it under… like, low-key, you know?

RG: Yeah, but not against the government—against *Linda*!

DF: *[Laughs]* Exactly. She's *worse*!

RG: That might be true. By the way, I was on the bus in Santa Monica the other day and I passed by the greatest bar. The name of the bar was The Bitter Redhead.

DF: *[Laughs]* Phil Dick's friend, two names, and, uh, 150. Minnesota. Outside of Minneapolis. But call these numbers back, or I'll call you on my other phone card.

RG: You mean the two numbers you already gave me?

DF: Yeah. 'Cause it might work now.

RG: I'll try.

DF: Or, uh, another one's 952-469-3726.

RG: Okay.

DF: Okay, I'll talk to you in a minute.

RG: Okay.

DF: Okay, bye.

RG: Bye.

[END]

DF: Uh, you called those numbers, right?

RG: Yeah, it didn't work.

DF: Yeah, I could hear 'em. They make the, uh, hang-up sound when you call. I pick it up and there's, like, you know, there's just blank... nothin'.

RG: Huh.

DF: And it goes, "Please hold one moment."

RG: Yeah, 'cause it actually rings, and then nothing happens.

DF: That's bizarre. And there's like numbers on the phones and shit, so I don't get that at all.

RG: The mechanical woman will say, "Thank you," and then there's nothing.

DF: Uh... okay, is that cool, man, so I can get the fuckin' the hell out of Minnesota? 'Cause I'm just worried because of the "license revoked" thing, that I get down to Wisconsin, or I just get someplace where, you know, the car's not in threat of gettin' taken from me. *[Pause]* Hello?

RG: Yeah, yeah, yeah.

DF: Okay... are you gonna do that tomorrow at some point?

RG: Sure. I was... I was *really* hoping that Minnesota would be a place where you could...

DF: Me too. This is insanity. I don't know what to fuckin' do. Can you look up the asylum stuff and, like... And I've gone through the stuff and checked the emails... th-the addresses of, like, the Fakespace and, like, all that stuff and read through all of them. Scary shit, eh?

RG: Oh, yeah, yeah. In fact, I was just showing the file the other day to Joan and, you know, Danny Sheehan's got a copy of it, and Randy.[13]

---

13. Just before Dion fled San Diego, I mailed copies of my complete file regarding the entire "Night Vision Goggle Affair" to my friend Randy Koppang, author of the book *Camouflage Through Limited Disclosure*, who then passed it on to his friend, attorney Daniel Sheehan, founder of The Christic Institute. Sheehan is perhaps most famous for having successfully represented Karen Silkwood in court against the Kerr-McGee nuclear power company, a case that was later made into the film *Silkwood*.

DF: That's really good, man. And… yeah, a class action suit's not out of, like, the realm of possibility… and winning money. I mean, it's like, fuck, man, I could go round up plenty of people in San Diego, uh… that… and it's funny, 'cause, uh… you know, fuck, you don't have to tell anybody that you're on speed, or that you had done it in the past… but the shit that they have out there, and that they're… like, the Navy… 'cause the last two places that were busted were on Miramar. Go to methvalley.com and there's a meth lab locator, and you can put San Diego in and it'll show you where the meth labs that were busted are, and they're on the military bases out there. There's no secret behind that because biohazard teams have to go out and clean this shit up. And they were busted on the military— *on*, not *near*, but *on* the military bases. But the shit sucks, though. It's not very good. *[Laughs]*

RG: *[Laughs]*

DF: It's not the good old kind that makes you fuckin' crazy.

RG: Well, I was telling Joan, it seems to me that they're purposely targeting areas where they know that there's junkies—

DF: Oh, yeah, of course. No credibility.

RG: Yeah, no credibility. And no ties, usually.

DF: Yeah… I can't believe this shit. I'm so far fucking away. I've driven *so far*, Robby, to get away from that crap. And it's just, like, *bam*, worse than before.

RG: Yeah.

DF: Escorted out of fucking Minneapolis for the most part. Yeah, I'm reading through, like, the *Snitch Culture* book and it's all, like… fuck, dude, you've gotta… I'm not gonna have any ability to call you. Let me make sure where I'm at, okay?

RG: Okay.
[Pause]

DF: Hello?

RG: Yeah.

DF: It's Lakeville, Minnesota.

RG: Okay. Lakeville.

DF: Yeah. And it sucks 'cause the location that I'm at doesn't have, fuckin', uh... uh... a Western Union. I don't know where the fuck... I'm gonna have to figure that shit out. What time is it there?

RG: Right now it's 1:25.

DF: So it's like 3:30 in the fuckin' morning.

RG: By the way, you know you just have to keep hanging in there because things are progressing, but extremely slowly...

DF: Yeah, I'm almost at the edge of, like, *suicide*... and they keep fucking pushing me to it. Thank you. I appreciate everything. It's the only way I'm going to be vindicated, man.

RG: I can't... even though it's progressing slowly, I can't get it to go any faster without—

DF: I know, and it's fucking lawyers and the justice system, and I'd rather be doing this on the outside than be locked up for something I didn't do. I really thought they were gonna plant something in my vehicle tonight. That's why I need to get the fuck out of here, you know?

RG: Well, yeah, we haven't even got to level zero yet. We're, like, at negative thirty. And we're climbing...

DF: Yeah. Exactly. And, like, man... if anything happens... are you still recording this?

RG: Yeah.

DF: *If anything happens to me* it is because Lita Johnston and the military and these people that are behind this *did it*. If I get killed in some way it is because *they did it*. I thought it was gonna happen tonight. I'm scared of these people. They're fucked up. They're menacing. They've threatened

me. They've tortured me. This is insanity. I do not under-
stand this at all. I didn't do anything. And it's just, like…
I'm scared for my life. And I'm glad that… like, I mean…
I'm glad I was able to drop off all the stuff I could to my
son. And I mean that should be obvious to anybody that
I'm in fear for my life, you know? And that's just, like,
fuck, man… I've got these phone cards… how's Eric doin'?
[Laughs]

RG: Eric's worse off than you are. It's called secondary
post-traumatic stress syndrome.

DF: *[Laughs]*

RG: Or actually it's tertiary. Because secondary would be *me*.

DF: Yeah, no… oh, it's funny. He's frettin' over it. *[Segues into
an Edward G. Robinson impersonation.]* "I'm big, see! The
bigger I am the harder you fall, see?" Oh my God. Fuckin'
ridiculous, man. I'm tryin' to, like, sit down and write and
do stuff, but this makes it a little difficult.

RG: *[Laughs]*

DF: I-I don't… I don't fuckin' get it, man.

RG: Well, you know Huey P. Newton wrote that book in jail,
right? *Revolutionary Suicide.*

DF: *[Laughs]* Yeah, my grandfather said he didn't write that
at all.

RG: Yeah, his tailor wrote it.

DF: *[Laughs]* Oh my God. Fuck that. I don't think I… I do
not want to be writing any more books from inside of jail
again. I still have all my stuff. Oh, it's weird, though! The
way they tossed up the van was—obviously they weren't
looking for drugs. But they did, like, abscond with my, like,
shaving stuff and my deodorant and everything. I couldn't
seem to find it anywhere in my van.

RG: That's weird.

DF: Yeah, it sucks because they tossed… I have this little, like,
doll thing that's like a Raiders fan that's in all black with a

hood? And it says… you know, it's got the Raiders symbol on the front and it says, SPOOKY. *[Laughs]* He's like this crazy little doll that's symbolic of a Raiders fan. Apparently one of the cops was not a Raiders fan because he hucked it all the way across, and stepped on it in the van.

RG: *[Laughs]*

DF: 'Cause all the other weird crap I got up on the dash was completely left alone… except for that thing was, like, all the way in the back. God damn fuckin' funny shit. You got my message from earlier though, right?

RG: What message?

DF: When I called from Peru's house?

RG: Oh, yeah, yeah.

DF: Yeah, that's what I thought. Okay, cool.

RG: That was pretty… it's funny how they literally, like, implied you were a terrorist without using the word because I guess they—

DF: Oh, yeah. "He's been talking to people about *doing* something."

RG: *[Laughs]* Which could mean *anything*.

DF: *[Laughs]* I know. No, but, all in the tone of voice.

RG: Right. It's all in the tone—yeah, sure.

DF: And the way that they were treating Jawbone. "Yeah, he's been talking to people." *[Segues into Edward G. Robinson impersonation.]* "Yeah, about doin' somethin', see? Talkin' to people, see?"

RG: *[Laughs]*

DF: "He's gonna be doin' somethin', see? He's got big plans, see?"

RG: Did anybody talk to Sunset or Weasel?

DF: No, I went back and asked. See, the thing is—oh, it's funny, 'cause there were about twenty-five of them sitting all out in front of her house after all that shit happened.

RG: *Twenty-five?*

DF: Man, no, there were about six cars all sitting out in front of, like… that I knew were, like, people of the *spook* variety. All down there. It's funny, 'cause I pulled up right next to 'em and just, like, look over. And I'm in this big, menacing van that has no muffler at this point—

RG: *[Laughs]*

DF: —oh, and this is another reason I need money is because the muffler has become unhooked at the engine, and the engine is right beneath me—*[Laughs]*

RG: *[Laughs]*

DF: —in the car and all the fumes are coming straight up through the bottom, and I'm surprised I'm not dead yet.

RG: Are you driving the vehicle kind of like the Flintstones?

DF: *[Laughs]* No, I'm *green.* I'm some sickly color and the van's filled with smoke. *[Laughs]* Because of carbon monoxide.

RG: *[Laughs]*

DF: It's just not a healthy situation at all. But it's only gonna take… and it sounds like a fucking 747's taking off with this muffler. *Bupp-bupp-bupp-bupp-bupp-bupp*! Oh, it's just so insane.

RG: Well, it's good you've got a stealth vehicle.

DF: *[Laughs]* I told my dad I was going to paint it fluorescent orange.

RG: *[Laughs]*

DF: He's like, "Yeah, you might as well."

RG: *[Laughs]*

DF: Yeah, and I'll know, like, after… I won't call for a little bit, and if I do… because I'm not sure if it's through the phone calls or if there's actually some fucking microchip shoved up my ass.

RG: Randy thought—and Joan thought too—that there might be some kind of tracking device in your vehicle.

DF: That's what I'm thinkin', but I've got to figure that out for sure by not giving you any more—

RG: Right, yeah, that'd probably be a good idea.

DF: Yeah, but, you know, and it's like I'm... 'cause I'm fuckin' *lonely* as a motherfucker, man... but, uh, I'll see by just not telling you anyway where I'm at, 'cause I specifically gave you guys—you and my dad—addresses.

RG: Yeah. It could also be the time—even if you don't give me any specific information—the amount of time you spend on the phone might have something to do with it.

DF: Yeah, 'cause these things... each time I've been on these weird phones there's been weird shit going on with them, and the book *Snitch Culture* talks about roaming wire taps the night after the 9/11 attacks and all that crap they have. And my dad, who's, you know, a cop—law and order—is like, "Yeah, man, they're just fucking up because they're gonna get all the laws they wanted to track terrorists taken away from them because of this. It's totally unproductive to what they want to do." And he's like, "They're idiots."

RG: Well, yeah, the pendulum's either gonna have to swing back real hard, like post-Watergate, or it's gonna get worse. I mean, it's one or the other.

DF: It's getting worse before it gets better. My dad, coming from, like, that total other side, goes, "Man, what they're trying to do, it's just gonna completely ruin them." And he goes, "Yeah, the higher up you go the stupider they get."

RG: *[Laughs]*

DF: And he's coming from the perspective of a cop. And it's like, I just gotta talk to him, "Look, man, you're dealing with something that's on a need-to-know basis, and you're at the bottom of the food chain here, motherfucker." He's gonna find out what's going on when they run my name tomorrow.

RG: Oh, okay.

DF: That's what he needs that information for.

RG: Oh, good, okay. Well, yeah, I sent that—

DF: That's enough of that because—

RG: Yeah, yeah, yeah.

DF: —I don't want it to change—

RG: Yeah, okay.

DF: —before tomorrow. And 'cause I called him tonight and all that crap, but he's been like—he told me, "I can't do that." And you know, I knew what he was saying when he said that and, like, everything... so hopefully... maybe my sister can find out as well and shit, but, uh... is there anybody... *[Pause]* We've got a minute left, man.

RG: Oh, okay.

DF: So... so what time're you gonna do that tomorrow?

RG: Sometime around noon.

DF: Noon your time?

RG: Yeah.

DF: Okay, can you *please* try... because that's two this time and I'm, like, in the middle of a parking lot and I won't be able to go anywhere until then.

RG: I'll try to do it a little bit earlier.

DF: Okay. J.P.

RG: Yeah.

DF: And, uh... okay. And that's it, just...

RG: All one word.

DF: All one word. Okay. Cool, man. Uh... how've you been? How's the Freemasons? How's the Brotherhood?

RG: *[Laughs]* The Brotherhood's been going fine. I've gotten a lot of interesting information from the Scottish Rite Library.

DF: That's—

[Line cuts off]
[Long pause]
[Then dial tone]
[END]

* * *

Three days later Dion called back. Here's how that conversation went:

CONVERSATIONS WITH DION, PART 2
(3-13-04)

DF: —just now with the, you know, migraine headaches, total nauseousness. There was—you wouldn't believe the amount of fucking people that were following me out of Duluth. It was disturbed. Just total complete insanity and absurdity. Dude, if that kid took something it was not those night vision goggles. This is just, like, insanity. Hey, Robby?

RG: Yeah?

DF: Uh… it's, you know… talk to me, man. I need to find… have you found anything on the fuckin' Internet or whatever for, like, countermeasures to protect me from that shit? And, uh, whatever else, to just, like, throw it back at 'em? 'Cause… you know, I mean, I'll talk to that lawyer on Monday or whatever, but this has just fuckin' got to stop.

RG: You know, I haven't even asked Joan about that because it hasn't been prominent on my mind since you left San Diego. I mean, you haven't mentioned it since then, so how do you—?

DF: Fuck, no. It's continuous. It's never stopped.

RG: Really?

DF: I just… you know, I mean, fuck, I've suffered from migraines since I was nine. It's not like, you know… I'm used to torture. It just got real bad up in Duluth. This has never even stopped. *[Pause]* Um… what've you been up to? I mean, fuck, didn't you get my messages yesterday?

RG: Yeah. I got your message yesterday. I tried to call you back, but... I called the number, and a guy answered the phone. I said, "Is Dion there?" And he said, "You have the wrong number," so I don't know—

DF: What?

RG: I tried to call the phone booth, wherever you were at, and I just... the guy said I had a wrong number, so...

DF: Well, I don't know. I mean, that was just a crazy—I was—I was—oh, I was in Lake Superior, Wisconsin. I was able to, like, circle around—it was like out of a movie, man. I'd circle around, go pull into a Wal-Mart or a Target, a fuckin' shopping center, and, like, trap 'em. And, I mean, there was this one dumbshit that I trapped inside of a friggin' car lot at, like, ten o'clock at night. You know. And this woman's in this big ass truck in a car lot. I circle around and I trap her in there. And I go run up to her car and bang on her car like, "Who do you work for? Why're you following me?" Some kind of a *fuckin'* movie, man. You know, and I'm banging on her car and she's like, "I don't know, I don't know," but she doesn't call the cops.

RG: *[Laughs]*

DF: You know, she doesn't... I'm like, fuckin'... then she's, like, honking her horn, trying to get me to move. And I'm, like... dude, there's cars passing, *bam bam bam. Nobody* calls the cops. And shit. And I sit there for a minute, and then finally I just... And, uh, just before that, this guy, like... I took pictures. I've got about fifty pictures, if not more, of, like, all these people, man. Here's my... here's my notes. They were tryin' everything to prevent me from getting down here, man. Guess what happened... my tire went flat. So, it's like, you know, I had to fuckin' run down to a mechanic to get a wrench. The whole nine yards of just ridiculous bullshit, you know? Just complete sabotage. And just everything possible to drive somebody fuckin' completely insane. I was seriously contemplating suicide. And stuff. Now at this point I'm just gonna get one of the motherfuckers out and beat their ass.

RG: *[Laughs]*

DF: You know. Seriously. 'Cause the military doesn't give a fuck about these snitches. They don't care about them at all, and I'm just gonna pummel one of the dumb motherfuckers until they… you know, have to go to the hospital. 'Cause I'm over it. I just did three days at Labor Ready, and… just, like, oh my God, it was such a mess. Torturous work. I did seven hours of work for fuckin' thirty-five dollars.

RG: *[Laughs]*

DF: Isn't that insane? And I'm talkin' about serious work. The last day the owner gave me forty bucks, then he gave me the Labor Ready, which was twenty-one dollars.

RG: Is that normal?

DF: Huh?

RG: Is that normal? That's how Labor Ready works?

DF: Well, they pay you minimum wage. Labor Ready gets… they'll charge the employer. So they're making half. So the employers are totally shocked when somebody comes in and does a really shitty job.

RG: *[Laughs]*

DF: It sucks, you know. It's 'cause Labor Ready's getting *half* the money. And they're not doing any of the fucking work. All they're doing is setting you up with a job. It's just total bullshit. It sucks. Labor Ready's particularly bad. Dude, you wouldn't believe the extent of this shit, and like how many people, like… You know about when I was talking to that guy Ben at that liquor store, right?

RG: Yeah.

DF: He came in and talked about the midgets?

RG: Yeah, yeah… he immediately said, "One midget came in after you, then another one came in."

DF: Yeah. Back to back. I mean, how often do you see a fucking midget?

RG: *[Laughs]* I've *never* seen one.

DF: Exactly. You wouldn't believe the extent to this, Robby. How, like… how many people are involved in this. It's insane. This is just, like, beyond… I mean, this is a vast network of people. Oh, I bought eggs! I was throwing eggs at their cars and shit when they passed by me.

RG: *[Laughs]*

DF: And stuff. And just like—

RG: Can you give a rough estimate as to how many people?

DF: Man, from San Diego to here? *Thousands* are involved in this. *Thousands.* Up in… jeez, up in, uh… Duluth and Lake Superior I was followed by over a hundred people. Literally.

RG: So when you go into each town they must call out the local people, right?

DF: Oh my God, it was so insane. There was a grocery store I went into. And there must have been thirty people in this grocery store that were there and were observing me and talking shit. I pull into a gas station and they've said stuff. They've laughed. They've made total fucking fools out of themselves and everything. It's just totally ridiculous, you know.

RG: And that Ben guy totally noticed it, right?

DF: Yeah. He's like, he's like, "Oh my God, I can't believe how many people just come in." And I told the guy at Labor Ready. I pointed out the predator drone that's followed me from San Diego all the way across the country.[14] I go,

---

14. This is an excerpt from an article titled "Stealth Army" (oddly enough, however, the table of contents page identifies the title as "Invisible Army") by Mark Thompson, published originally in the Jan. 9, 2012 edition of *Time* magazine:

"The hot military trend is something the Department of Defense isn't exactly known for: economy. It's a new era, indeed. Today's generals and

"Check that thing out." He goes, "Yeah, man." I go, "That's not a star, is it?" He goes, "No, that's way too bright." I go, "That's the predator drone that's followed me all the way across the country." He's like, "Oh my God, man." And me and him... oh! Oh, dude, you know your smoking invisibility gun?

RG: Yeah.

DF: Go onto fucking wired.com and look at "Marines: Hiding a Few Good Men."

RG: Okay.

DF: It's the military application of that. I read the article, but on the computer at Labor Ready I couldn't see the pictures.

RG: Oh, that's too bad.

DF: Yeah, so go on to that. And I need that whole dossier overnighted to that lawyer on Monday... or if you can do it as soon as possible so he gets it on Monday. Because I took all the money from working these three days into the van and it's running real well. I'm headed back that way.

---

admirals want weapons that are smaller, remote-controlled and bristling with intelligence.

"In short, more drones that can tightly target terrorists, deliver larger payloads and are some of the best spies the U.S. has ever produced [...].

"Drones had a big year in 2011, and their stock continues to rise because another big buzzword is *persistence*, the ability of a drone or satellite to simply stare at a target until something noteworthy happens [...].

"So scientists and engineers in skunkworks across the country are busy honing drone 2.0 technology, which will be able to deliver bigger payloads and operate with greater stealth—which is a reminder that any conversation about the wisdom of relying on these weapons always lags the technology that makes them possible.

"America's arsenal has become so small and lethal, you don't need the U.S. Army—or any military service at all, in fact—to field and wield them. The CIA, which used to be limited to derringers and exploding cigars, is now not very secretly flying drones. With little public acknowledgement and minimal congressional oversight, these clandestine warriors have killed some 2,000 people identified as terrorists lurking in shadows around the globe since 9/11. Expect the tally to go higher this year."

RG: Remember, you have to request the free consultation, and then mention Harriet.

DF: Well, fuckin' A, I need to write all this down, 'cause I can't find my fuckin' notebook now. Also, you got pen and paper?

RG: Yeah.

DF: Oh, well, you're recording—

RG: I'm recording it.

DF: —so you can just play it back. Okay, so... shit... uh... oh, Home Depot has, like, free phones that can be dialed out of, but I'm trying to, like, work on that. My dad says he's gonna put more money on that phone card, 'cause this has been insane. I thought I was gonna get killed at that Starter Photo and shit. Because, I mean, I fuckin' ran out there and was beating on the woman's car. *"Who do you work for? Why're you following me? Who sent you?"* It sounded, you know, it sounded like, fuckin'... how many *movies* has that been in?

RG: *[Laughs]* Right.

DF: You know? I mean, fuck it. I'm, like, after I got done with it I'm walkin' away goin', "What the fuck did I just *say?*" How insane is it for somebody, like, to be beating on someone's hood going, *"Who do you work for?"* It was just ridiculous. I need someone to start a website for me called "free-dionfullerfund.org."

RG: Oh, actually that's a good idea. We could just put all the documents up there. Oh, make sure... yeah, tell the lawyer you've got photographs and everything. In fact, bring the camera in with you.

DF: I'm holding on to all that.

RG: Yeah, okay.

DF: I still need to get the friggin' address from you too.

RG: Of the lawyer? You don't have it?

DF: I can't find it. But here's some of the other... you know how people got that "SUPPORT THE TROOPS" stuff on their lawns? I'm gonna make a sticker that you put right underneath it that says, "BUT FUCK OUR LYING, THIEVING GOVERNMENT."

RG: *[Laughs]*

DF: I mean, I'm just now getting the back of my van organized so it's livable. They've been treating me like a mother-fuckin' *animal.* This is insane, Robby. I want such fuckin' revenge. I'm gonna hurt one of these motherfuckin' people if they continue to do what they're fucking doing to me. I'm trying to clean out my van and the fuckin' caps or the fillings in the back of my teeth are rattling around in my fuckin' head, man. They had this thing that was directed at me where I could hear... imagine, like, you know, hitting a hammer on concrete, and that's what it sounded like when my teeth were hitting each other. Continual migraines, nauseousness... just, like... did you get on to that website I told you... "force-.com," I think it is?

RG: Yeah, I punched in "force-dash," but I couldn't find—"

DF: You didn't get it?

RG: No.

DF: 'Cause it shows all the stuff and, like, how you can buy this shit. And I need somebody to... fuckin', I just want to arm myself with all this crap. Because it's *available.* All this shit that they have is totally available to the average citizen now. You know, because that's who it's made by. And for a relatively cheap price. But you'd crack up at what they have, and how *cheap* it is. There's just shit that, like, when someone pisses you off, you just want to have, you know?

RG: Yeah.

DF: Like, there's this thing... really, really small thing that makes this insanely loud, irritating noise that you could just put somewhere and, uh, every five minutes it does that. It makes this loud, irritating, "Doodle-loo," and people will stop whatever the fuck they're doing and rip everything off the wall.

RG: *[Laughs]*

DF: I'd go all over the country and do that, man. You could just... oh, it's just ridiculous what you could do. It's so cheap. But, uh... don't forget the "freedionfullerfund."

RG: That's an excellent idea.

DF: Thank you. 'Cause it's, like, man, they're fucking driving me god damn insane. Oh, while cleaning out my van I saw somebody had set their office cubicle outside near a trash-can that I was using and I wrote in big letters in magic marker, "HARASSING INNOCENT PEOPLE, BAD. CATCHING BAD GUYS, GOOD. GET IT RIGHT, YOU NAZI RETARDS!"

RG: *[Laughs]*

DF: Uh... it's fuckin'... you can't imagine, like... it's scary. There's gonna be a war in this country. Because of the amount of people that those guys have. Nazis convinced that they're doing something patriotic. They're just fucking Nazis.

RG: Oh, yeah, well, they've all been told that you're Timothy McVeigh or something.

DF: Yeah, exactly. And they're just doing their... just like the fuckin' Nazis, they're following orders. And it's funny, 'cause the rhetoric's not even different. Because instead of saying they're following orders now, they're just doing their *job*.

RG: *[Laughs]*

DF: That's how they rationalize it, you know. I mean, that guy in that bathroom at that gas station was, you know, he's like, "Man, just give the shit back." I mean, he felt bad for me. 'Cause he'd saw what he'd done. But you know he just goes home and looks in the mirror and goes, "Man, I'm just doin' my job." Just like the fuckin' Nazis, man.[15]

RG: So when that guy came in and said that to you in the bathroom, he sounded sympathetic?

DF: Yeah. What he said was, "Just give the stuff back and it'll all stop." You know, he sounded sympathetic. But then right when he said that, *bam*... they don't even let their own people be alone for very long.

RG: Right. Yeah, they're as trapped as anybody else.[16]

---

15. This incident deserves special attention. Upon first being escorted out of Minnesota by the police, Dion strolled into a bathroom at a rest stop to clean up a bit. A guy, some normal-looking Joe, walked in after him. The guy took a piss at the urinal, then walked over to the sink beside the one in which Dion was washing his face. As the guy cleaned his hands, staring at Dion's reflection in the mirror, he said quietly, "Just give the stuff back and it'll all stop." A few seconds later someone else entered the bathroom, at which point Dion's new friend walked out of the bathroom quickly—almost as if the dude had gotten spooked by the presence of this new stranger. Dion didn't even finish washing up. He got back into his van and hightailed it out of there.

16. On Raven1's website, various victims of the gang stalking project would often post messages on the site, including advice on how to deal with the trauma. Raven1 requested that the harassers *themselves* post messages. Most of the resultant messages seemed to be total bullshit—fourteen-year-old kids imagining what it would be like to be a badass government agent. The only message that came across as truly genuine to me was this one:

"I wanted to be in a position of control. The feeling of having so many lives at my fingertips is just a rush. I used to run a small arms deals business outside of San Francisco, until the Feds finally had the evidence they needed to storm my building. Specifically, my prints on stolen goods.

"But I was able to elude them for so long, and they were impressed with the organization I was able to build, so they made me a deal. Federal prison, or work for them. It doesn't take a rocket scientist to figure that one out.

"At first, the control that I was able to extend to those assigned to me was a rush. Every time I made a decision that could make or ruin a life, I just got a run of adrenaline. The feeling was astounding. That feeling only lasted for the first few months. My superiors have since then tightened our restraints to ensure that we can't contact the outside world. They insist that no one know we exist and anyone who knew me before had thought I had died when I was recruited. This conspiracy goes much deeper than you think it does.

DF: Yeah. And there's *thousands* of them. It's insane. It sucked…
when I was in Duluth and Lake Superior, there's nothing
but bars and strip clubs and shit. I couldn't even do anything.
I couldn't even stop at all to enjoy myself. To do *anything*. To
spend five dollars at a fucking bar and call you from a strip
club. You know. That's all I fucking wanted to do. And the
amount of technology and shit that they had directed at my
fucking head was vibrating the mirrors on my car. Imagine
that. Imagine what that's doing to my fucking skull. What's
it doing to me? You know. That can't be good.

"The government is little more than a front for an even large [sic] organization, and those people are ruthless.

"More ruthless than I or anyone I work with. They could just as easily sentence a room of 5-year olds to death as they could shower in the morning. Hypnosis isn't even the beginning of it, although we have used hypnotic techniques to make some forget of our existence, we mostly operate through covert operations. Terrorism, seemingly illegal acts.

"Consider this: when you see a news story about a man who killed his neighbor and was arrested, that neighbor is most likely one of the people my superiors doesn't want to exist, and the person who did it is one of our operatives, and I can guarantee that that person never went to jail for a second, rather, he was brought back to [the] nearest mobile command center.

"We are paid well, there is no question about that, but with all the security measures we must go through, there's nowhere we can spend it. I can't even go to Wendy's without risking being recognized, just like many other of my co operatives. So our money mostly goes to purchasing supplies and ordering food and things from places that will deliver food or that we have an agreement with.

"I only have about 5 minutes left to type before the counter-IC programs go down, so I'm going to have to make this quick. You must know about the new static field generator that my agency is developing. It distorts all light and sound waves that pass through it creating an illusion much like invisibility coupled with a dead spot of noise. In other words, you won't be able to hear or see whatever is in the radius of the static field.

"The thing returns false radio, UV, IR, and thermal signals, as well.

"We've field tested small models that can cover a single individual, and have had no real problems with those, but soon we will be developing larger ones to hide aircraft and large artillery."

RG: Are they projecting it at you from a car that they're following you with?

DF: That's what I can't... well, that, yeah, definitely. But, uh, Eric had some good advice with the headphones and the music and shit. That seems to, like, help a lot when I'm driving, you know?[17]

RG: Oh, really?

DF: Yeah, when I was fuckin' in the grocery store... oh, I finally got the muffler and stuff. The car's running so much better and, uh... and, uh... what was I gonna say? So I fixed that and, like, a couple of other things and, like... just able to put a little bit of fuckin' money back into that thing, so it keeps running. In Carlton, Minnesota, at the mechanics', I had to get a mechanic to take the break pads off my fuckin' car 'cause they were kinda jacked up so while they were doing that I went to the truck stop to take a shower. I go over to the truck stop, and I don't know if it's a coincidence or what—my luck is just, like, so insane, it's *not* coincidence—the water pressure on the shower is gone. So it's the first time in, like, a week and a half I'm able to take a shower, and I'm in there, like, enjoying it, thinking, "Oh my God, this is great. I'm just going to stand here and enjoy this one simple fucking pleasure." The water pressure goes.

RG: *[Laughs]*

DF: Then it starts to spurt a little bit again. Then it goes again. To the point where I have to get out of the shower. So I get out of the shower and I shave and do all this other shit and get back in the shower. It ends up being a miserable *fucking* experience. I go over to get my car from the

---

17. For some reason, headphones and Sonic Youth albums (perhaps atonal music in general?) seem to stave off the effects of nonlethal weaponry, at least according to Dion.

mechanic. There's one guy in there who *looks* like he works there and he goes, "What brings you out here?" Kind of, like, in a weird tone of voice. I go, "Well, you know, I have a son in Minneapolis," blah blah blah. Then this "young-ster"... nineteen or twenty years old... because I asked his age... he's like, "Man, what's up with, uh, you know..." He asks, but not in so many words, about the situation. I go, "Look, man, fuckin', this kid I knew for three days stole this shit off Camp Pendleton," and I tell him, like, the story. And he goes, "Man, you're a wreck." I fuckin' turn to him and go, "What the *fuck* did you say?" And he goes, "Uh... uh..." I go, "What the *fuck* did you say?" And he's like, "Oh, dude, don't beat me up." And I go, "What the *fuck're* you... man, you don't know me, you don't know anything about my motherfuckin' life and you better shut the fuck up." And he, like, walks away. I was gonna kill the guy. And he knew... about *something*... about me. This nineteen-year-old fuckin' dickhead. Then he comes back and says, "I don't want to get off on the wrong foot," and I go, "Dude, just fuckin'... you have no idea what's going on, and this could be *you*." And these people are so indoctri-nated that they just don't even get it. There's no empathy behind their eyes at all. It's scary, man. I don't know what the fuck is goin' on in this country, but it's almost to the point where it's out of control. Just off topic... listen to me. I sound like I'm on speed, don't I?

RG: *[Laughs]* Well...

DF: Don't I?

RG: Uh, yeah...

DF: People think that. And it's weird. It's something that's directed at me.

RG: Actually, you just sound agitated to me.

DF: Well, that's what it is. Is that what it is? It's funny, 'cause, like, this is what'll happen all the time. People don't really see it, themselves getting agitated too if they're, like, right in my area. Now, if they're in my field or area it'll happen to them too. I mean, it's pretty directly able to be focused on me and shit.

RG: I remember one time you said you were riding your bike down in San Diego and there was a halogen light that swept over you and made you dizzy. Are you still experiencing the dizziness that you described back in—?

DF: Oh, yeah, yeah. *[Reading from his notebook]* "Hitler didn't tell his people to be evil. Hitler told his people to be patriotic." Fuckin' dumbasses. *[Reading from his notebook again.]* "Nationhepatitus-c.org." Can you go on there and find a hepatitis C study for me to do that pays money? Uh, so I can get some kind of… There are these real good books called *Cointelpro* by Nelson Blackstock, *Police State America: U.S. Military "Civil Disturbance" Planning* edited by Tom Burghardt, and this one looks *really* good: *Dreamer of the Day* by Kevin Coogan.

RG: Yeah, Dave Emory's interviewed Coogan a billion times on his show.

DF: That totally ties into all this shit. It's weird, because these people have been indoctrinated and they don't even fucking realize what they're doing. They're just following orders. Just doing their job.

RG: That's the biography of Francis Parker Yockey, the guy who wrote *Imperium*.

DF: Really?

RG: Yeah. That's *Dreamer of the Day*.

DF: Jeez, okay. Uh… *[Reading from his notebook again]* "Tell the truth and run."

RG: *[Laughs]*

DF: *[Still reading]* "He who is happy is hidden." Uh…

RG: Do you need me to give you the address?

DF: Yeah, hold on a sec. Um, yeah. Don't let me go without getting that. *[Still reading]* "Labor Ready has taught me that poverty is the great, forced integration."

RG: *[Laughs]*

DF: It is, man.

RG: I'm gonna go get the notebook with the address in it. Hold on… *[Pause]* Hello?

DF: Yeah.

RG: Okay, here it is.

DF: Wait, hold on a second. Oh, guess what the cops went through when they searched my van. They just went through all my paperwork.

RG: Really?

DF: Paperwork, yeah. Robby, be careful. This situation with the… I'm tellin' you right now, if this situation with the lawyer gets any kind of weird… 'cause I literally… this is *vast*, and I don't know who's been compromised. And, like, these people's names that you have… that you've talked to, like this woman… be *very* careful with who you trust.

RG: I—

DF: You know, that lawyer woman you talk to—

RG: I think she… no, I mean, I trust her.

DF: See, that's, like… be very careful. Because what these people are… a lot of them are informants. And in the past they've gotten involved in stupid, fucked up situations where now there's pressure being applied to them. Some of them are just, like, you know, full-on military brats. Just little dickheads. But then there's other people who have the total ring of sincerity about them… but, man, you can tell that something… some kind of, like, blackmail—I'm pretty sure—is being applied to them to make them do shit that, you know, they might not necessarily, like, agree with? You know what I mean?

RG: Right.

DF: It's like the saying goes, "If it's too good to be true, it probably... you know... it probably is." So, like, if this situation with this fuckin' lawyer doesn't go very well tomorrow—on Monday, I mean—I'm on the fuckin' outlaw-punk-rock-stolen-gas-tour fuckin' back to San Diego to blow some motherfuckers up. No, that's a joke, obviously. Your phone's tapped and all this other shit. And it's, like, I know you're being... trying your hardest and working... man, you, like, saved my life. I would've fuckin' offed myself a long time ago if it weren't for you and my pops helping me out, you know? 'Cause, like, everything that's happened over this short period of time is just ridiculous, you know? I mean I almost fucking killed myself in San Diego. I *tried*. You would talk to me... you know, like that night on the phone... just to repeat for this tape recording's sake. These people almost *killed* me. There for the grace of God I walks amongst us. *[Laughs]* Funny.

RG: By the way, Joan's also trying to find private detectives in Minnesota that might help as well. However, the reason she's given me James Anderson's name is because he's actually represented a lot of people who were victims of, say, ritual abuse and—

DF: Oh, really? Okay, great! Because this—yeah, okay, that reminds me. Where's my fuckin' notebook? There's the weirdest shit, like, disappearing. I know there's something in the car that's giving out a signal because I've driven out to areas that are so fucking out of the way, like... oh, man, before I forget, there are homing crows that have followed me all the way from San Francisco... I mean, uh... San Francisco? That's where I want to go to. San Diego. There are fucking crows that have literally followed me all the way... fuckin'... across the country. I know it sounds crazy. But they've been able to do that with pigeons and stuff. I mean, they're able to train birds.

That's just something that's been going on since the dawn of Man, and the military… the applications… the military have always used birds. It's insane seeing these two crows just, like—they'll cross in front of my car. Up near Lake Superior they were real prevalent. I'm gonna buy a BB gun.

RG: *[Laughs]*

DF: No, I am. And just, like, *off* the motherfuckers. I need to get a pellet gun that's, like, capable of popping tires too, so I can just walk up to these people's cars and just pop their tires, y'know?

RG: Do they look like real crows?

DF: What's that?

RG: Do they look like real crows?

DF: Dude, they're real crows.[18]

---

18. Perhaps not. Consider the following information, as reported in the June 2011 issue of *Popular Science* (pp. 34-35) by science writer Joshua Saul:

"In 2006, Darpa, the Department of Defense's R&D arm, commissioned AeroVironment, a company specializing in remote aircraft, to create an unmanned aerial vehicle (UAV) small enough to fly through an open window. AeroVironment had already built the 4.5-foot-wingspan Raven, which first saw combat over Afghanistan in 2003, but making a UAV so much smaller took five years and 300 different wing designs. Finally, AeroVironment has a working prototype: the 6.5-inch-wingspan Nano Hummingbird. "It was never our intention to copy what nature has done; it's just too daunting," says Matt Keennon, the UAV's head researcher. The camera-equipped bird beats its wings 20 times a second, whereas hummingbirds clock up to 80. Still, it can hover like the real thing, plus perform rolls and even backflips. Here's how the bird flies.

"WINGS: A skeleton of hollow carbon-fiber rods is wrapped in fiber mesh and coated in a polyvinyl fluoride.

"CAMERA: The camera angle is defined by the pitch of the Nano's body. Forward motion gives the operator a view of the ground, aiding navigation. Hovering is good for surveying rooms.

"BODY: It weighs 18.7 grams (less than an AA battery). The craft is remote-controlled, but an onboard computer corrects speed and pitch.

"TO FLY: By beating its wings back and forth, the UAV creates lift by deflecting air downward, creating an area of high pressure directly below the wings and low pressure above.

RG: Okay.

DF: No... the thing is, like... you know how loud a crow is? *[Imitates the screech of the crows]* It's also got a psychological warfare aspect. It's irritating as *shit. [Laughs]*

RG: *[Laughs]*

DF: That's another thing I need you to look into—is, like, psychological warfare. Because that's, like, the only tool to my advantage. These people have handheld mikes. So they're, like, little things in their hands. And they're such idiots that when they're passing me they'll, like, brush their hands up to their face real quick to go, like, "Well, he's going northbound now." They done it all over the place in San Diego. Adam pointed them out to me, like, "Watch, man." And it's, like... it's really stupid. I told you about it in the past. It's, like, jeez, they should've just used cell phones or something because it looks more natural. But I've gotta ask my dad about it as well. And if you have any questions about practical applications of surveillance and tailing or any of this crap, try emailing my dad questions about that stuff, you know?

RG: Okay.

DF: Uh, 'cause, like, go... when I'm telling you something try to get, like... email him to get some kind of verification or whatever to get the stuff, uh... so I can figure out what the patterns are and whatnot to get, fuckin', *revenge*. Uh, and there's, like... also, type that into a search engine, "revenge," and it'll come back with the funniest fuckin'

"TO TURN: Increased angles mean more thrust. If the Nano Hummingbird sharpens the angle of its right wing on each forward stroke, and does the opposite on each backstroke, the craft rotates clockwise.

"TO MOVE FORWARD: The wings beat symmetrically. If the angle at the end of the forestroke and beginning of the backstroke on both wings decreases, the nose dips downward and the craft moves forward.

"TO ROLL: By increasing the angle of only its left wing, the Nano Hummingbird creates more upward thrust on its left side, which will cause the bird to roll to the right."

things, man. There's... there's *millions* of websites that're specifically designed for revenge. It's great.

RG: *[Laughs]*

DF: You'll get the funniest fuckin' stuff you could possibly imagine. It's hilarious. There's a lot of really pissed off people on this planet, man. And it just gets scary, dude. I don't even, like... you know, I mean, I'm worried. You know, you look at it and you're, like, "God damn, people actually sit around thinking this shit up?"

RG: I hope Joe Malone's not reading any of those websites.[19]

DF: *[Laughs]* I know! Oh my God, that's who's behind all this.

RG: *[Laughs]* Well... let me give you the address.

DF: Yeah.

RG: You got a pen?

DF: Yeah.

RG: Okay. It's, uh... James Anderson.

DF: James... Anderson?

RG: Yeah. E-1000 1st National Bank Building. *[Pause]* 332 Minnesota Street. *[Pause]*

DF: Okay.

RG: St. Paul. *[Pause]* 55101 is the zip code.

DF: I don't even need that.

RG: Okay. Here's the phone. 651-227-____.

DF: Okay.

RG: And remember to—do you remember the woman's name?

DF: No, I need to write that down.

RG: It's Harriet McDougal, which is spelled M-c-D-o-u—

DF: M-c-D... *[Pause]* Go ahead... I don't want to say it out loud.

RG: M-c-D-o-u-g-a-l.

DF: D—say it again?

---

19. Joe Malone was a suspected child molesting serial killer Dion and I worked for when we were both seventeen. True story. Alas, it's too complicated to go into now...

RG: M-c—

DF: No, I got that. When I say, when I stop, and then when I go, "D-o-u," just finish it.

RG: Okay. "g-a-l."

DF: Okay. Sorry, that's, fuckin', for the people in the background. Okay, cool. Uh… so what's been up with you, man?

RG: *[Laughs]* Well, I've been talking to—

DF: Can you turn that off now?

RG: Oh, okay.

[END]

14.

According to Dion, James Anderson essentially believed his story but told him that no one had ever successfully sued the U.S. military. It would be a losing battle, and lawyers don't like to fight losing battles. They prefer to fight battles they know they're going to win beforehand.

So from Minnesota Dion went to Kansas, where he somehow ended up being abducted by two rednecks in a village called Winona on March 18, 2004. Unfortunately, I didn't have the recorder ready for that particular phone call, but I began transcribing it only seconds after I hung up the receiver:

* * *

The phone rings at around eight o'clock. I pick it up.

"Hello?" I say.

A guy with a thick Southern accent says, "You Robert Guffey?"

"Yeah."

"You've got a friend named Dion Fuller?"

"Yeah."

"We've got a situation here."

In the background, I can vaguely hear Dion arguing with someone else, some guy with an equally thick Southern accent. Has Dion been arrested again? Is this some overzealous sheriff and his deputy looking for a reason to rough up the local vagabond?

"This friend of yours is a real storyteller," the dude says. "Claims he's bein' chased by invisible midgets and government agents and even weirder things. Abe here ain't buyin' any of it. Your friend says you're the guy who can confirm it all. Is that so?"

In the background the argument seems to be heating up dramatically.

The receiver is ripped from the first dude's hand. "*Hello?*" this new voice says. I assume this is Abe.

"Yeah?"

"The demon liar Satan has taken over the body of your friend. He speaks with the tongue of Lucifer himself! Have no fear, brother, we have the ability to rid your friend of this evil spirit fiend. Zack's heatin' up the tools right now. It won't be long before your friend's a new man again."

In the background, I can hear Dion's voice growing even more stressed.

"Right," I say. "Okay. That sounds rational. However, allow me to inform you that Dion's not lying at all. You see, back in July a friend of a friend went AWOL from Camp Pendleton and ended up crashing at his place. The dude stole a DOD laptop computer, a 9 mm Iraqi gun, and twenty-five pairs of night vision goggles. When the Navy tracked down the equipment, they arrested everybody who was staying there at the time, including Dion. Well, Dion didn't have anything to do with it, but the Navy didn't believe him. They thought he was part of this vast smuggling ring operating out of Camp Pendleton, right? So they throw him in jail and interrogate the shit out of him, but after six days they know they don't have any hardcore evidence against him. So they release him. That was back on July 24th of last year. But ever since then they haven't let him out of their sight for even a second. They've been torturing him with electromagnetic devices, tracking him, harassing him, blocking his path every step of the way. I've experienced all of this right along with him since last July. It's been going on for about, oh… nine months, I guess?"

After a slight pause, the dude says, "*Wellllllllllll…* the government's done even worse things in the past, I guess."

"That's right. Look at Watergate, the Gulf of Tonkin, Agent Orange, Waco, the list goes on and on."

"What do you do?"

"I'm an English professor at CSU Long Beach." I don't mention the fact that I've been unemployed for three months.

"Hey, Zack," the guy says, talking away from the phone, "hold off on those tools for a minute." An image flashes in my mind: one of those banjo-playing mutants from *Deliverance* heating up branding irons on the kitchen stove—instruments to eject the spirit of Lucifer from an unwitting human vessel. Abe puts his lips back up to the receiver. "You know, this government's getting out of control. Everybody's a damn criminal now. It's damn near Luciferian!"

"That's exactly right," I say. "And it's gotten even worse since 9/11. They're keeping a close eye on everybody who doesn't fit into their neat tidy little system."

"It's all about control, brother. Have you noticed the little tricks Hollywood tries to play on you? They try to make every damn abnormal thing in the book seem normal. Now they're shovin' homo-*sexuality* down all our throats. Can't turn on the TV without seein' some homo gettin' married on the news and the comedy shows try to make it all seem so funny."

"Right," I say. "*Queer Eye for the Straight Guy. Will & Grace.* It's rampant."

"It's sin, that's what it is, it's sin. Pure and simple."

"I live out here in California, and I know people who work in the *Industry*, and you'd be amazed at what goes on. There's this place called Bohemian Grove up near Sacramento where all the major politicians and movie actors and foreign royalty gather every July 15th to enact this ancient druidic ritual called 'The Killing of Care.' They worship a giant forty-five-foot stone owl. There's a camp in the Grove called the North Woods where people like Henry Kissinger and Gerald Ford have sex with high-priced hookers. This journalist named Alex Jones actually snuck in and managed to videotape the ritual on a digital camera. You can see little snippets of it on a website called infowars.com."

"Well, I don't got no Internet connection."

"Maybe you can go down to the library."

"Yeah, maybe I'll do that…"

"I know it sounds unbelievable but—"

"No, no, I believe ya. I mean, hell, that's nothin' new. Back in them old days the early Catholics would always absolve themselves of sin by having *sessssual* intercourse with Temple prostitutes! Apparently they're still doin' the same thing today! The Catholic Church is nothin' more than a pagan religion. Not many people out here know that them there Catholic holidays are just old pagan holidays all trussed up with new names 'n labels 'n stuff."

"That's absolutely right," I say, "December 24th isn't even Jesus' birthday. That was the birthday of Nimrod, a druid, one of the so-called 'wise men' who actually ended up marrying his mother. That doesn't sound too 'wise' to me. Yeah?"

"Ay-men, brother! Nimrod's wife was the Moon Goddess, and Nimrod was a *sssorcerer*, heavy into *ses* magic! They made December 24th Jesus' birthday just so they could keep all that pagan stuff around without anybody knowin'. Jesus was really born in October. Say, how do *you* know all this stuff, anyway?"

"I have a scholarly interest in religion. I'm currently researching the Pagan roots of Catholicism."

"Then you better take a look at a book called *The Two Babylons* by Alexander Hislop. That goes into all this stuff. *All* of it! This damn black magic stuff is everywhere, so I ain't surprised it's over there in Hollywood too."

"Do you know what the word 'Hollywood' means?" I say, turning toward my computer and typing the title of Hislop's book at the top of a blank screen.

"What?"

"In the ancient days druids used to gather in secret groves to cast spells with these wands they thought were capable of casting magic spells—illusions. The wands were made out of a special type of wood. Holly. Hollywood. So it makes sense that the modern day druids, filmmakers, should choose 'Hollywood' as the name of their own special little grove where they cast spells on the rest of us."

A moment of silence. For a second I'm thinking I've lost him. Then he yells, "Spells! To fuck with our minds!"

"That's exactly right. Mind control."

"Have you ever read *The Protocols of the Elders of Zion*?"

An anti-Semitic hoax favored by the Nazis. Now I know exactly where my new friend is coming from. "Sure, of course. Essential reading. In *Behold a Pale Horse*, William Cooper says that every time the word 'Jew' appears in the book you should replace it with the word 'Illuminati.' Have you ever read *Behold a Pale Horse*?"[20]

"No." This surprises me. "I've heard of the Illuminati, though." This does not.

---

20. An underground bestseller when initially published in the United States in 1991. The book postulates a covert alliance among the United States military, extraterrestrial forces, and ancient secret societies for the express purpose of destroying democracy.

"You should read it. You can learn a lot from that book. Cooper died just recently. The cops shot him in Arizona."

"Do you know John Singer?"

I pause. I think about it, wondering if it's a name I should know. "No," I finally say.

"I knew John Singer's wife. Back in the '80s the Mormons murdered John Singer. The cops said he shot himself, but that's a damn lie. *A damn lie!*"

"I'd like to know more about that, actually. You know there's a strong connection between Mormonism and Freemasonry, of course."

"No… no, I didn't know that."

"Oh, sure. Joseph Smith, that's the guy who founded the Mormon Church, was a Freemason himself and just stole all the rituals and stuff and transplanted them over to Mormonism. That's why the Masons had to kill him. You never saw the Mormon Temple in Salt Lake City? The whole building's ringed with pentagrams."

I think for sure I'm telling him stuff he already knows. Instead he says, "Whew. I just got a chill up my spine. Hell, I didn't know that."

"Sure. Go check out utlm.org, this rogue Mormon website, and you can see photos of the pentagrams in the windows of the Temple. Pretty blatant."

I hear him writing something down. "I'll be sure to do that as soon as I can get to the library."

"So did I manage to confirm everything?"

"Well, you damn sure as hell did, mister!"

"Did Zack put the tools away?"

"They're coolin' off." He moves the receiver from his lips again. "Hey," he tells somebody in the background, "untie him, okay? It's all right. Yeah, yeah, I'm sure of it. It's fine." He presses his lips to the receiver once again. "Can I have your phone number?"

"Sure. 310-320-7035."

"And what's your name again?"

"Robert Guffey."

I can hear him writing it down. "You want to talk to your friend now?"

"Sure."

I hear the receiver being put down on a hard surface, then some voices in the background, about three of them. Then I hear Dion say, "Hello?"

I try to sound casual. "Hey, how's it goin'."

"You just saved my life."

"I know."

"Listen, can you Western Union me some money? The van broke down on the side of the highway where these guys picked me up—somewhat against my will. I need it fixed and I need to put some gas in the tank. I need to get back to California, man." He sounds worried.

"Then you can hang out with us druids at the Grove."

"What?"

In the background I hear Zack yell out, "Hey, that thing's still out there! It's gotten bigger!"

"Here," Dion says to Zack, "tell him what you're seeing."

"Wait a minute, I don't understand," Zack says as he takes the receiver back from Dion, "if this guy already knows what's goin' on, why do I have to confirm anything for him?"

"No, no," Dion says, "you're not confirming anything for *him*. He's got a lot of people around him who don't believe any of this is happening, so the more witnesses we have to this stuff the better."

"Well," Zack says to me, "professor," he coughs nervously, "sir," he coughs again, "all it'd take is fifteen minutes drivin' around with the guy and you'd know somethin' weird's goin' on. I don't know what it is, but it's definitely somethin' weird. Not too long after I pick Dion up off the side of the road this light comes rarin' down out of the sky, right? I'm talkin' lower than cloud cover here. Weren't no star. No, sir. And it just starts followin' us, y'know? Followed us all the way back to our house. Hell, it's outside right now. What is it, some demonic UFO kinda thing?"

"That's the predator drone," I hear Dion say in the background, "it's been hounding me ever since San Diego. Whenever I leave the van it picks up my tail somehow."

"I can see it too!" Abe yells in the background. "I'm gonna take a Polaroid of this thing. What's all this damn fool nonsense about invisible midgets?"

"Oh, let my friend explain it," Dion says.

"Tell Abe," I say, "that the military has this—"

"Wait, wait," Zack says, and hands the receiver over to Abe.

"Yeah, yeah, go ahead," Abe says, breathless with anticipation.

I say, "Apparently the military has this technology that's like a suit, right? And it makes the person wearing it blend in perfectly with his surroundings. It's more like hyper-camouflage than invisibility. They have really small people wearing the things. That way they can get in and out of tight spaces fast. We don't know if they're using midgets exactly—that's just kind of a humorous exaggeration on our part. But if I had technology like that, I'd definitely give it to very thin, agile people, like dancers or acrobats."

"Or demons," Zack says.

"Yeah, or demons. Sure. Demonology is a wide open field. A lot of different entities could be included in that category. Apparently, the U.S. military is testing this type of technology—and similar things, like electromagnetic non-lethal weapons—on junkies and small time crooks in San Diego. Y'know, just people on the fringes with no ties to anybody else. Check out an article called 'Being Invisible' on wired.com. It'll tell you more about it."

"This's gotta stop. I'm gonna talk to my pastor about all this."

"Yep. Good idea. Maybe he'll be able to do something about it. Can I talk to Dion again?"

"Sure."

"Oh, by the way, thanks for helping my friend out."

"Well, I don't know if I *have* helped him any—not until he chooses to accept the pure light of Jesus into his heart."

"Listen, don't worry about it. Even if that never happens, I'm sure you've managed to help him. That can only give you brownie points on the other side, y'know? Say, can you maybe get him some gas and make sure the van's running smoothly? I don't want to put you out any but it'd be a shame for those military guys to suddenly—"

"We can do that, professor, sir, we certainly can."

"Thank you."

"Here's your friend again."

"Thank you."

Dion takes the phone back. "Hey."

"I just thanked him for helping you out. He said he wasn't sure if he did or not."

"Oh, no, you did," Dion says to Zack, "yeah, yeah, you really did," then mutters into the phone, "Man, if you could only see this guy. Jesus."

"I don't want to."

"I'm starin' at a copy of *The Protocols of the Elders of Zion* sittin' on his bookshelf right now."

"I was surprised he'd never heard of *Behold a Pale Horse*."

"Yeah, that's what I thought was goin' on in their heads too," Dion said. "I recommended they read it, though."

"That's good. Introduce some culture into their lives."

Dion laughs. "Listen, I need to get back to California as soon as possible, man."

"Don't worry, it's all set. Your two buddies there are gonna fix the van and put gas in it. So I'll see you when I see you. If you get here by next Wednesday we can go see *Mr. Arkadin* at The Egyptian Theater. They're doin' a whole Orson Welles retrospective right now."

"That's cool. But if we go, y'know, about two dozen midgets are gonna be followin' right behind us."

"Great. The Egyptian Theater needs some more business. Of course, the invisible ones won't even have to pay. They'll just walk right in. I hope we don't sit on them accidentally. Hell, that'll really confuse them, when you drive all the way from Minnesota to Los Angeles just to see a fucking Orson Welles film. The next thing you know, The Egyptian Theater will end up on the Terrorist Watch List."

"I'm glad you can laugh about all this."

"Who's laughing?"

"I did it," Abe yells out. "I got it on Polaroid!"

"I think I have to go back to the Bo 'n' Jeb Show now," Dion says.

"Pray they don't whip out their banjos."

"I'm worried about the branding irons, not their banjos."

"See ya."

"Later."

Dion hangs up.

15.

Bo 'n' Jeb managed to get Dion a job working in a wheat silo in Winona, Kansas. With the proceeds from that job he was able to buy a house. Yes, an entire god damn house! Apparently, in 2004 the economy in Kansas was way better than it was in California. I was still looking for a new job. The problem with teaching college is that if you miss out on getting any classes at the beginning of the semester, you can't just pick up a teaching gig somewhere else. The semesters start at the same time everywhere. So, unless I found a job at a bookstore or some awful place like that, I'd have to wait until the fall before I could get a teaching gig again. Since I was collecting unemployment off the classes I taught the previous semester, I decided to visit Seattle (where my ex-girlfriend Stephanie had lived for years). I planned to take my resume to various universities and community colleges with the hopes of moving up there permanently in August. Another reason for visiting Seattle was to try and shake off the feeling of permanent paranoia that Dion had been instilling in my psyche since the previous July.

Meanwhile, I had been talking to Dion's ex-girlfriend Melanie on the phone more and more frequently—at first just to give her updates on Dion's misadventures, then later because I enjoyed talking to her. She had stored up a cornucopia of misadventures *almost* as insane as Dion's (but not quite). When I said I would be visiting Seattle soon, she suddenly told me she had just dumped King Louie and planned to move to Seattle sometime during the summer to visit her parents on Whidbey Island and to get over the lingering trauma of the break-up, so maybe we should meet up.

I said we could do that.

From March to June Dion remained in the house in Winona. The house was in the middle of nowhere. He'd set up a lookout area on top of the roof where he could see anything or anyone coming from miles around. Since moving to Winona all his problems had ended. The little flying saucer seemed to have disappeared (in fact, the last night he'd seen it was the first night he met Bo 'n' Jeb). No mysterious strangers had threatened him in a public restroom. The invisible midgets weren't tossing his furniture around anymore. He had a steady job in the silo, he was gaining all the weight he'd lost in San Diego, and everything generally seemed to be looking up for him. We concluded that Winona was so damn small it would be impossible to engage in the street theatre/gang stalking tactics that are so easy to pull off in major urban centers like San Diego or Los Angeles or Austin. There's so much chaos going on all the time in cities like that, who would notice or care if the meth addict down the street was being driven hi-diddle-diddle off the deep end by a bunch of government flaks? Any new resident in Winona would be noticed by every single person in town. You couldn't just roll a bunch of jarheads into Winona and blend in with the crowd. Everyone in the crowd knew each other. Also, people in small towns own guns and if they hear something prowling around their house they tend to shoot first and ask questions later. Invisibility's fine for practical jokes and fraternity pranks, but it won't stop a bullet from penetrating a skull. I wonder how many invisible midgets had to die before that simple fact was fully appreciated. No doubt the lab boys must still be working the bugs out of that system.

As the weeks turned to months and Dion grew healthier, and more relaxed, I noticed that (conversely) he seemed to grow more tense. Believe it or not, Winona, Kansas, is a very boring place. I could tell Dion was getting antsy. As much as he despised his time spent as a lab rat in an urban cage being assaulted on a regular basis with a lethal arsenal of virtual reality Boris Vallejo paintings, at least it was exciting. You never knew what was going to happen next. Now he was spending his days moving around bales of wheat in a silo waiting for a tornado to take him away to Oz. As Hunter S. Thompson once said, "Being shot out of a cannon will always be better than being squeezed out of a tube." Despite his protests to the contrary, I always had the distinct feeling that Dion secretly missed being incarcerated in prison. Everything's much easier there, after all, and you don't have to make stupid decisions on your own. The stupid decisions are made for you.

Sooner or later, I suspected, Dion would try to make his own tornado rather than wait around for one to find him.

Unless, of course, he had some reason to stay where he was. I believe he came across a stray cat and took that in, to make up for what he lost in the desert. He was always trying to call his ex-girlfriend Jessica. He'd write her crazy letters offering her co-ownership of the house if she'd come live with him. I don't think most twenty-four-year-old girls would be too enticed by a house in Winona, Kansas, co-owned by a man who's on the run from angry invisible midgets and roving metal eyes. Jessica, alas, was like most twenty-four year-old girls. She never responded to the letters.

16.

In June I hopped on a plane and headed to Seattle, feeling relief that I was leaving the Dion situation behind at long last. It was over. No more craziness for me. I was going to Seattle to relax. I didn't even leave a phone number for Dion. I figured he could get along without me for awhile.

The day after I arrived in Seattle I received a phone call from Dion's ex-girlfriend telling me she was in town and that we should hook up. I gave her Stephanie's address, and her parents dropped her off. That night Melanie took me on a whirlwind tour of Seattle that included the infamous Seattle landmark The Lusty Lady (an all-women-owned-and-operated strip club where Melanie had worked a few years earlier), a gay bar on 13th Avenue named The Cuff, and a club called Neumos in Capitol Hill where we saw a band called The Catheters. All the band members were good friends with Melanie's ex, King Louie. At one time Melanie had been friends with everyone in the band but now wanted to beat up one of the musicians' girlfriends for something this girl had said about her on the Internet. While taking the bus to the club Melanie just kept ranting about this girl: "That bitch is gonna be surprised when she sees my fist coming at her face, particularly since she thinks I'm in New Orleans right now. She can't get away with talkin' shit about me on some board and think there's no consequences just because it's the interfuckin'net. Fuck that whore! Listen, Robert, you better be ready to hold her down while I kick her in her cute face with my boots. That bitch! That stinking cunt bag! No one's gonna want to fuck her after I get through with her, certainly not the fuckin' bass player in the fuckin' Catheters. She's gonna have to find a new face and a new fuckin' occupation. She ain't gonna be the jism receptacle for *that* band no more, no way. Stupid slutbag."

She reminded me exactly of Dion. She even laughed like him and had the same screwed-up sense of humor. She was pretty much Dion with a vagina. She also had his propensity for talking shit, and then not following through on it. When we reached Neumos at last, Melanie somehow found the girl she was looking for in that teeming crowd, and they immediately hugged and acted like they were lifelong friends. No mention was made of the supposedly offensive comment that had been posted on the Internet for all to see.

Melanie and I didn't return home until almost sunup. We spent the night in Stephanie's apartment and woke late in the afternoon. One of the first things I did upon awakening was check my email on Stephanie's laptop. Dion had sent me a message asking me where I was. I made the mistake of telling him I was in Seattle and had just met Melanie for the first time.

I couldn't predict the reaction this email would provoke in Dion.

The next day I received an email telling me he had boarded up the house in Winona, thrown all his belongings (except for the new cat) into the back of the van, and was now on his way to meet up with us in Seattle.

hey,

i'm leaving Kansas—I registered the van—Insurance, etc. I'm finally legal i've disassembled an entire chevy van engine and its in the back of my van—just in case...
no more Western Union, Robby! now You can give me cash in person!!!

See You Soon
Dion

I should've just kept my mouth shut. The very nightmare I was running away from was now headed toward me in a black windowless death van.

This same day Stephanie kicked me and Melanie out of her apartment for reasons I still don't understand, so at the last minute we were forced to find a motel room to stay in for the night. The next morning we responded to an ad in the newspaper and found a place for Melanie to live (she hated staying with her parents)—a room on the second floor of a ramshackle house east of Broadway on Capitol Hill, not far from Seattle University. She would be sharing the house with two other girls in their twenties, one of them a local DJ named, of all things, Super Jew. I spent that entire day and the next helping Melanie move all of her belongings from her parents' house on Whidbey Island to this little room in Seattle. Melanie's parents were very pleased to see me and treated me like a king, probably due to the fact that every other male they had ever seen her with was a lot like Dion. They kept making very surreal comments, as if they thought I would be marrying her soon.

To my surprise, after we'd finished moving everything into the room, Melanie told me that I really *would* make an ideal husband. I wasn't sure what the hell was going on, as she didn't seem to be joking. I began having disturbing flashbacks to the previous summer when I spent most of my free time having pointless arguments with the Insanely Jealous Poet about the premature subject of unholy matrimony, so I took off the next morning and saw a double feature of *Godzilla* (the original one made in 1954) and *Fahrenheit 9/11* at the Varsity Theatre not far from the University of Washington. By the time I returned, sometime around 4:00 P.M., I accidentally walked in on Melanie fucking some dude I'd never seen before. He looked like a scrawny, acne-ridden teenager. Apparently she'd picked him up from a dive bar down the street on Broadway (or so I learned later). Melanie seemed mortified, so I apologized for interrupting their festivities, then wandered over to Broadway and grabbed a slice of pizza at Pagliacci's, a joint I used to frequent when I was going to school in Seattle back in 1996. Seeing how many junkies were wandering up and down scenic Broadway, I realized this place would be Heaven for Dion. I jotted down the following description in my notebook:

The Man in the Armani Suit stands on the corner of Broadway and Pine with his hand held out, asking for change. He isn't begging, he's asking. Quite polite. Not pushy at all. Well spoken. Patient as a glacier.

And like a glacier, the man seems cool and aloof, unaffected by his surroundings. Broadway and Pine lies smack-dab in the middle of Capitol Hill. On one corner stands the modern, Lego-like exterior of Seattle Community College. Right across the street: The Egyptian, a Masonic Temple that was converted into a movie theatre fifty years before. The Masonic symbols (the square and compass surrounding the letter G) are still emblazoned in the ornate stained glass windows circling the old 1930s-style building. Junkies in their late teens and early twenties, all of them Caucasian, track marks standing out on their pale skinny arms like miniature landing strips, roam up and down the street begging for change. Despite their youth, they look empty and haggard. Purplish half-circles frame their eyes. Shadows pool in their sunken cheeks. A pixie-like girl with short auburn hair calls out, "Kiss my ass for a dollar, kiss my ass for a dollar!" An old man in an incongruous Hawaiian shirt pauses in front of her for a moment, as if considering the proposition. Before he can decide, however, the Man in the Armani Suit (standing only a few feet away) reaches into his coat pocket, pulls out a handful of pennies, and hands them to the girl.

"Here's eleven cents," the man says, "keep your ass."

"Hey, thanks," the girl says, sarcasm further tainting her already exhausted, raspy voice, "now I can move to fuckin' Tahiti."

The Man in the Armani Suit just stares at her for a moment. "Do you even know where Tahiti is?"

The girl shrugs, as if to say, "Who cares?"

"Listen, I gave you that eleven cents so you'd go away," the man says. "This is my *corner. You're siphoning off my possible income."*

The girl stares at him, her head slightly tilted to one side, as if seeing him for the first time. "Hey, why the fuck're you even here?*" she says.*

I put my pencil down and thought, Jesus, Dion's gonna have to come up with a pretty good shtick to compete with these ass-clowns. I had confidence he could do it, but didn't even bother predicting how the shtick would manifest itself this time.

After I finished my pepperoni and feta cheese pizza, I took a bus back to Stephanie's apartment. I had to explain to Stephanie that I now had no other place to stay. On top of that I gave her address to Dion, which meant The Man Who Came to Dinner would no doubt be sinking his roots down into her carpet very soon. Exasperated, Stephanie nonetheless allowed me to sleep on her couch that night. Dion arrived the next day, but I was at a party on the other side of town and wasn't there when he aggressively tried to manhandle Stephanie's breasts in her kitchen. Stephanie was used to Dion's drunken she-nanigans and didn't take the fumbling attempt too seriously. After shutting down his overtures, she pretended nothing had even happened, then assured Dion he could park his death van along the curb outside her apartment.

This was the first time I was seeing Dion in person since all the nonsense with the goggles had begun almost a year before. He had changed a great deal. He'd grown a nasty moustache that made him look rather like Chester the Molester—and the death van wasn't helping his image any. He took me out to the van and showed me the back of it, which looked like the Punisher Mobile in those old Marvel comic books I used to read back in high school. He had everything back there one would need to carry on a fruitful existence, including home-made weapons, canned foods, a hot plate, paperback novels, a copy of the U.S. Constitution, a complete set of Tony Robbins subliminal self-help CDs, and hardcore porn. He also showed me a photograph of what he looked like a few days after leav-ing San Diego. The photograph was incredible. He looked like an Auschwitz victim—or worse. He looked like a scarecrow who'd simultaneously contracted cancer, leukemia, *and* AIDS. In the photograph there were even patches of hair falling out of his scalp.

He now sported a healthy, thick head of hair, as he'd always had before the nonsense in Pacific Beach. I was convinced that his apartment must have been flooded with intense doses of some kind of radiation—either electromagnetic energy or microwaves—for a long time, perhaps going all the way back to July. He said he felt much better since settling down in the house in Winona, and his urine wasn't red anymore.

"What did you do with your house?" I said. "Did you just lock it up?"

"Well," Dion said, "the guy I bought it from suddenly told me he wanted it back. He was claiming some kind of fraud or something. He claimed I wasn't giving him the money I owed him, but he's friends with all the cops in town and I was an outsider so I knew the cards were stacked against me, so fuck it. That's bullshit. I don't like getting ripped off. So I just filled all the plumbing with wet cement, then locked the door and took off."

"I guess you're not going back then."

"Fuck no. Winona's for chumps. Where's Melanie? I want to see that bitch."

"The last time I saw her she was fucking some teenage kid. I don't even know where he came from. I accidentally walked in on them. I only left her for a few hours…"

Dion laughed and shook his head. "Same old Melanie. Does she know I'm coming?"

"I mentioned it." One of the reasons Melanie wanted to get a new place quick was so Dion wouldn't be able to find her.

"Where is she now?"

Thankfully, I've always been bad with directions, so he believed me when I said I couldn't remember the exact address. I said it was somewhere near Broadway. Melanie had begged me not to reveal the address to Dion. "If he knows where I live he'll never leave," she said, "and then I'll be doing heroin with him again and it's all gonna go downhill. I can't *live* like that anymore!"

So I just kept mum (or tried to).

It was about two o'clock in the morning when I returned home from the party, so by the time Dion and I finished catching up the sun was rising over Seattle.

"This is the first time I've ever been to Seattle," Dion said. "I like it here. I think maybe I'll stay."

Melanie and Stephanie will be glad to hear that, I thought. "Great," I said, "as long as you didn't bring any invisible midgets with you." We then exchanged some clever invisible midget jokes. Right before I went back inside Stephanie's apartment to go to sleep, I asked Dion if he was carrying any drugs on him.

He said no. "I don't want to do that shit anymore," he said. "I've been clean since I left San Diego and I want to keep it that way."

I said that was an excellent idea, bid him good night, then went into Stephanie's apartment to go to sleep at last. Dion crawled into the back of his death van and did the same.

Stephanie woke me up around noon to tell me that Melanie was on the phone and wanted to talk to me. So I wiped the sleep from my eyes and accepted the call. Melanie wanted to know if Dion was in town. I said yes. She then asked me to come over, but *not* to bring Dion with me because she didn't want to start doing heroin again.

"He says he doesn't have any of that shit on him," I said.

"He's lying," she said. "Can you come over? I'm hoping you'll spend the night with me. But don't bring Dion with you. *Please* don't. Trust me on this."

"But... where am I supposed to say I'm going?"

"Don't tell him anything. Just get over here. *Please.*" I wasn't sure why it was so important that I go there immediately. This "vacation" I had planned was quickly descending into a mundane variety of intrigue that was now driving me up the wall more than the fucking invisible midgets ever had. Once again I promised her I wouldn't give Dion her address, then hung up.

I glanced out the window. Dion's van was no longer parked along the curb. Since I had no way of getting ahold of him, I figured there was nothing I could do except go ahead and take the bus over to Melanie's house.

This journey took a couple of buses and several hours.

By the time I arrived, Dion and Melanie were sitting on the front porch drinking beer and laughing with each other. Jesus Christ, I thought, I could've gotten a ride here with Dion!

"I looked for you right before I left but you were gone," I said to Dion. "Where were you?"

"Just sightseeing," he said. "Then I remembered that you said Melanie lived off Broadway, so I just wandered up and down the streets until I saw her out here on the porch."

Melanie had a frozen look on her face.

We sat on the porch for several hours as Dion regaled us with the several scrapes with death he'd endured on the way here from Winona. So far, he had not experienced any of the harassment that had been so prevalent on the journey to Winona. I watched Dion and Melanie barrel through several 40s followed by bottles of Peppermint Schnapps and Grey Goose. At around 6:00 P.M., as the sun was just beginning to set, the teenage kid Melanie had picked up the other day suddenly arrived and sat in the chair next to me. That's when I learned his name was Noel. He was nineteen. Noel put his arm around Melanie and acted as if she had been his wife for several decades. Melanie didn't seem to mind this at all. Since it's not terribly exciting to watch three other people get drunk while you yourself are sober, to while away the time I pulled a small notebook out of my back pocket and wrote down everything Noel said over the course of the next few hours. Here's what I wrote:

1 CONVERSATION, 47 EPIPHANIES

Scrabble for your breakfast.

I didn't have shit, man.

I never really got punked out in jail.

The shower and the shitter's the same thing pretty much—nice dick there.

You're all "pshoo."

You had someone cut your hair in jail, dude. You get free haircuts in County.

I think the East Coast must be soft. Fuck that. I'm not tryin' to see Corcoran.

I did fuckin' thirty days for stealin' a car up here.

When I say bitch.

You want me to slap him around a bit.

You're fuckin'… your dad put you somewhere?

Cry for me please.

That's the first fuckin' gun I shot when I was twelve years old, .357 dude.

Monte Carlo. Sans hydraulics. I put a little juice in that motherfucker, dude.

I took out the 305 and god damn, fuckin' whore on wheels.

Keep the black guys in the neighborhoods.

I listen the fuck out of Art Bell.

Fuckin' cop's son in fuckin' prison. Better not let them know.

Buenas noches, senorita. That's Mexican for fuck off. Makes me feel smart.

What time is it, like fuckin' eleven o'clock?

Nice fuckin' smile there, buddy. Actually, I am nice. I'm a fuckin' pussy cat.

I was talkin' about my fuckin'… what's it called when you can't reach something?

Hey. It's not an open invitation, assfuck.

Fuck me running.

I'm also like a girl.

This guy stole my fuckin' shtick, man.

Can I be cool? Will you give me a dollar if I can be cool?

There should be no fuckin' talkin' about what he knows? That's cool. I'll hold him up on my shoulders.

We're havin' deep conversation about Eminem.

I don't think it's that complex. I think this guy's simple as shit. That's simple to fuckin' compute. I've got 18,000 different personalities. I'm not cool cuz I'm not sellin' records.

That's still him though. It's still him though. So he's got the Slim Shady bank account? He can't access the money until he's eighteen? He stole my whiteness. He made me black.

You fuckin' buy it. It's got a bar code on it. Fuck Slim Shady. Fuck his mom. Fuck 8 Mile. *His mom's got two pussy holes too, one in front, one in back.*

He's drawin' pictures of Marshall Mathers' cock. Can I turn this shit off now?

I'm in a fuckin' big butt suit. I'm fuckin' Tom Green. Eminem's not an ass, is he?

Jackin' off a moose is a fuckin' national holiday.

You got Irish drinking music? Why're you playin' Eminem, you queer?

Fuckin' Dropkick Murphys.

Are you gonna play the fuckin' CD or ya gonna masturbate on it?

Wanna go to the bar? How does it feel to want? I don't have any fuckin' money. I've got quarters. Does that count? I'm waitin' for your roommates to come home so I can fuck 'em.

Fuckin' molester moustache, dude.

No, he just drives around retirement communities. Fuckin' old ladies. Then I said, "Fuck you, old ladies." You takin' that down?

Grab my fuckin' beer before it gets yanked.

Fuck that. I don't want to do that shit.

Where you goin' there?

Both your hands have fuckin' alcohol in them. Why would you, man? You've got glasses for god's sake.

I've been here all year.

This guy's gonna burn your house down.

At around 11:00 P.M., Super Jew, who was a very attractive statuesque blonde in her early twenties, arrived home to discover three very drunk people and one bored writer sitting on the porch. Apparently Melanie had chosen to move into a house full of straight-edge devotees who abhorred alcohol in all its forms. Later Melanie told me she picked that particular house because she thought having straight-edge roommates would help curb her own drinking problem.

Super Jew grew more and more upset. At one point she asked Melanie how long the party was going to continue. Melanie couldn't respond through her laughter. At around midnight she summoned Melanie into the kitchen and asked her point blank if she were a prostitute. Melanie started laughing again. At around 1:00 A.M., Dion cornered Super Jew in the kitchen, grabbed her tits, and tried to have sex with her on the kitchen counter. Meanwhile, Melanie and Noel had gone upstairs, no doubt to fuck. Super Jew, who was about three feet taller than Dion, came storming out onto the porch, tossed Dion's limp body toward me as if he weighed no more than a blowup doll, and insisted I grab my friend and get the fuck out of the house. I wasn't quite done with the sentence I was writing at that moment, and tried to explain this to her, but she snatched the pencil out of my hand and acted as if she were going to stab my eye with it.

"You don't have to defend me," Dion said. "All I wanted her to do was go down on me. Man, this bitch really is a Super Jew, ain't she?"

Super Jew, for some reason, flew off the handle after hearing that and started screaming at us that was she was going to call the cops on the two of us, even though I wasn't doing anything except sitting there writing.

"Oh, okaaaay," Dion said, slurring his words, "it's jus' like a Super Jew to call the Super Stormtroopers." He started walking backward off the porch and almost fell on his ass. "Who're you gonna call next? The *Super* Anti-Palestinian Task Force?" Dion smiled to himself, as if proud he'd managed to think of that. "The Palestinians, *and* the people who live there, are an enslaved state. Two rights don't make a wrong, you cunt! *Two rights don't make a wrong!* Deep down, you know 'm right!"

Super Jew had already gone inside to call the cops. We could see her through the window picking up the phone.

I grabbed Dion's shoulders and said, "Let's just go. This could get bad."

"It's gonna get *worse!*" Dion said. He paused to whip out his pierced, malformed cock and pissed on Super Jew's front lawn. Then we walked around back to where Dion had parked his van. He crawled behind the steering while I got into the passenger seat. He rested his forehead on the wheel and started the van. He began to drive with his head still resting on the wheel.

There were no seat belts in the van. There was no lever on the inside passenger door. Once you were in, you couldn't open it to get back out. You could only exit through the driver's side, or wait for the driver to come around and open the door for you. There were plenty of other things wrong with the van. The engine made a sound like a skyscraper being excavated. Pedestrians on the street looked at us in fear. It sounded as if the van were going to explode or implode at any moment. Dark smoke sometimes spewed up *inside* the van, forcing you to roll down the window to breathe. It was a moving, thinking death van that seemed at times to possess a form of sentience all its own. How else to explain the fact that the vehicle seemed to drive itself while Dion appeared to be in a coma?

Just before we were about to get onto the freeway, I asked Dion to stop the car. I was thinking I could find a phone and ask Stephanie to come and pick me up (wherever the hell I was). He refused. "You're perfectly safe," he kept saying. "I've driven like this hundreds of times."

The only way I could convince him to stop was by telling him that *I* wanted to drive. He stopped on the side of the road, walked around the van, and opened the door, at which point

I jumped out and said I was going to call Stephanie from the gas station down the street.

"That's bullshit," he said. "You don't need to do that. Why make her drive all the way out here? It's just gonna piss her off. I can get you home. Or you can drive."

I knew that my sober driving was no better than Dion's drunk driving, so eventually he convinced me to get back into the van. To this day I still don't know how he managed to do that. Instead of getting on the freeway Dion decided he wanted to sightsee. Moments from that evening flash through my mind from time to time like genetic memories of World War II death marches. I recall that Dion somehow ended up driving up onto the sidewalk in downtown Seattle and continued to do so for a long time without killing anybody or getting arrested. Dion didn't seem to think there was anything unusual about this. The irony of me being killed in a van *I* bought for Dion escaped me at the time. I was too busy white-knuckling the dashboard and praying I made it back to Stephanie's apartment in one, human-shaped piece.

And then came the freeway. Dion was driving in two lanes at once. When we reached Stephanie's neighborhood I just kept repeating to myself, "We're almost there, we're almost there, we're almost there." Dion drove up onto the sidewalk once more and parked the van on Stephanie's lawn. Then he climbed rather gracefully out of the van, strolled around to the passenger seat, and opened the door for me.

"See?" he said. "That wasn't so bad."

My legs were shaking so much I almost fell down when my feet made contact with the earth.

"Okay... well... I guess I'll see you tomorrow," I said as I staggered toward Stephanie's apartment. "By the way, Stephanie said the cops around here might hassle you if you keep the van in one spot for more than a night. She said there's an isolated spot on the other side of that park, the one down the road, that might work out better for you."

He nodded, said he looked forward to seeing me tomorrow, then drove away.

I walked inside the apartment to find Stephanie sitting up watching TV. I told her everything that had happened.

"These people are insane!" I said. "I never should've come here. Everything was fine with me in Torrance and Dion in Winona. What did I unleash?"

Stephanie suggested we just wake up early and spend the day doing mundane, "normal" things. She suggested taking a tour of the underground city in downtown Seattle that had been destroyed in a massive fire back in 1889.

Compared to being attacked by Super Jew with a No. 2 Ticonderoga pencil and almost smashed to pieces on the I-5, that sounded good to me.

17.

Dion vanished for the next three days. Melanie kept calling, but I told Stephanie to tell her I wasn't around. "Tell her I died or I was abducted by flying monkeys, whatever." I didn't care what the excuse was. I just wanted to be left alone.

Near the end of the third day, Stephanie and I were watching the evening news. I was surprised by the fact that the lead stories on the local news shows were quite different from the ones you'd see in L.A. In L.A. the lead story was always something like, "Thirty-three children died today when their school bus was blown up by a disgruntled postal worker suffering from AIDS dementia." In Seattle the lead story would be something like, "Man steals ATM machine." That's it. It was almost disconcerting.

On this particular night the lead story was about someone the cops were calling "The Goldilocks Bandit." The newscaster reported that for the past three nights the Seattle PD had been overrun with complaints from one of the richest areas of Seattle. Someone was breaking into their homes in the wee hours of the morning, preparing a big meal in the kitchen, eating it while apparently watching TV (Nickelodeon), and then leaving without taking anything else. The thief liked ham and cheese sandwiches and Jell-O.

Stephanie and I just stared at each other.

"Do you think...?" Stephanie said.

"Who else would *do* that?" I said.

"What should we do?" she said.

"I'm no snitch," I said. "He's not going to break in here. I'm sure of it."

"Yeah?"

"Well, almost."

A day later Dion rolled up outside Stephanie's apartment. I crawled into the back of the van and asked him where he'd been. He said he was spending some time on Broadway with Melanie. She was looking for a new place to live ever since Super Jew had kicked her out. He told me he and Melanie were very excited to find the exact Jack in the Box parking lot where Kurt Cobain used to score heroin.

"Cool," I said. "By the way, are you the Goldilocks Bandit?"

"The what?"

"All over the news there're these reports about this guy who breaks into people's houses, has a little picnic in their living room, watches some TV, then bails without taking anything else. Is that you?"

Dion looked slightly nervous. "Fuck, why would I do that?"

"It sounds like something you would do."

"Fuck, if I'm gonna go to all that trouble I'd be stealin' more than just food."

"Speaking of which, Stephanie's been really fuckin' stingy with her food. Do you have any you can share?"

"Sure," Dion said, always the gracious host, spun around, and opened the tiny refrigerator in the corner. Back issues of *Buttman Magazine* and *Nugget* sat on top of the ice box. "What would you like?" Inside the refrigerator were stacks of orange and strawberry Jell-O.

I asked for a strawberry and he picked an orange one. As we sat in the back of his death van eating Jell-O I asked him what he planned to do in Seattle.

"I thought I'd live here for a while," he said. "Melanie promised to show me where the welfare office is tomorrow. Apparently you can get welfare in Seattle like *that*." He snapped his fingers. "It's the easiest state in the country to get welfare. Fuck, if I'd known that I would've come here a long time ago."

"Okay," I said. "But what about until then? It's probably going to take a long time for the paperwork to go through, right?"

He waved his hand in the air. "I'll just get a day laborer's job for a little while. I've done it before, I'll do it again. Besides, I've got plenty of money saved from that wheat silo job. Look." He pulled a shoe box out from beneath his bed and showed me wads of one hundred dollar bills. "I closed out my account right before I put the cement in the plumbing."

After I finished eating the Jell-O, I wished him luck and told him I had to go because me and Stephanie were planning on visiting the Space Needle today.

"All that touristy shit," he said, snorting with disgust.

"Yeah, it's pretty stupid," I said, "but Stephanie has her heart set on it."

"Can I come?"

"We really haven't spent that much time alone together since I got here," I said. "There's been all these interruptions. That was supposed to be the whole reason I was coming up here—to spend time with her. It hasn't worked out that way so far."

Dion nodded sadly, then said he was going to drive back to Broadway and try to find Melanie. She had already paid a month's rent, so Super Jew was stuck with her for a few more weeks.

Her and the Goldilocks Bandit.

* * *

Two days later Dion called Stephanie's house and asked me if he could borrow some money.

"Didn't you just show me a shoe box full of one hundred dollar bills two days ago?"

Dion hesitated, as if he hadn't remembered doing that. "Yeah, well... I lost it all. Someone broke into my van and stole it, man. Son of a bitch. That's what you get for trusting someone."

"Well, listen, you know I've been unemployed since January. Why do you think *I'd* have any money?"

"Well, how did you get here?"

"What does that matter? I'm collecting unemployment, plus I've got some extra money stored away—you know, extra money *besides* the money I already gave you for the van."

"I offered to put the van in your name. You can be co-owner."

As if I'd want to be co-owner of a van that would soon be found in the custody of the Goldilocks Bandit. Great. "No, thanks," I said. "It's a gift. Friends give each other gifts. But not today. I don't have any extra cash. Why don't you ask Melanie for the money?"

"Oh, I already borrowed money from her. I'll just have to think of something else," he said. "I'll talk to you later."

The next day reports hit the papers that the Goldilocks Bandit was now stealing cash and credit cards along with her porridge.

* * *

Melanie stopped by Stephanie's apartment the next day to pick up a purse she'd left behind when she spent the night. She took me into the other room, suddenly broke down crying, and admitted that she had done heroin with Dion in the back of his van. She even let him "borrow" some money to help pay for the heroin.

"I'm so stupid," she whimpered, "please stay at my apartment. You're a stabilizing influence. I know you'll prevent me from doing something stupid."

"What happened to Noel?" I said.

"Oh, Dion blew that whole thing. Noel was talkin' smack, so Dion beat the shit out of him and stole all his money."

"That Noel guy was a bit of an asshole," I said. It took me a while, but somehow I managed to calm her down and promised I'd be over later that night.

The second she left I locked the door behind her and said to Stephanie, "I'm hopping on a plane and going back home tomorrow. It's been a nice visit. If Melanie calls here please tell her I was killed in a tornado on the way back home."

"There aren't any tornados in the Pacific Northwest," she said.

I just shrugged.

* * *

The day I returned home I called the Chair of the English Department at Cerritos College and asked if they had openings for the fall. She said yes and scheduled an interview with me. A few minutes after hanging up the receiver the phone rang, so I figured it was the Chair calling me back for some reason.

But it was Dion: "Hey, when did you leave?"

"I got a job offer to teach summer school," I said, "so I had to come back to California. Sorry. I didn't know how to reach you."

"That's great that you've got a job. Say, listen, I know you said you didn't have any money, but if you could just Western Union me a few hundred dollars I swear my dad will pay you back."

"Your dad agreed to do that?"

"Yeah, I've already talked to him."

I sighed. Even if that were true, I wasn't really in a position to trust that his dad was going to come through for me this time. I had a limited amount of funds at my disposal at this point.

"What the fuck is going on?" I said. "Why do you need so much money so fast? You should've just stayed in Winona."

"I wish I had, my friend, I wish I had! The people here in Seattle, they fuckin' suck, man. No morals. No ethics. No nothin'. No one wants to help each other out. Everyone's out for *themselves* here, man. And there's a blatant prejudice against junkies."

"What're you talkin' about?"

"I was walking down Broadway and saw a sign posted in the window of a bookstore that said, 'NO JUNKIES ALLOWED. WE'RE TIRED OF FINDING BLOODY NEEDLES IN THE BATHROOM.'" I laughed. I'd seen the same sign. I thought it was pretty funny. "It's fuckin' *bullshit*, man!" he said. "I know people are prejudiced against junkies, but I've never seen it advertised so blatantly. It's a violation of one's basic civil rights, man."

"Dion, there's five thousand Kurt Cobain wannabes staggering up and down Broadway like zombies, panhandling and squeezing pus out of the sores on their feet and droppin' bloody needles behind them like molting feathers. Would you want to deal with that shit if you owned a business there?"

Dion grew indignant. "They *can't* discriminate against us! The whole economy of Washington would collapse in a day without us junkies! If not for us—!" He started rambling. What began as a series of simple logical fallacies soon fractured into a meaningless rant.

After a while I interrupted him and said, "Dion, you still haven't told me what you need the money for."

"Oh. I was gettin' to that. You see, the cops stole my van, man. They stole *your* van! The one you bought for me!"

"What do you mean they *stole* it?"

"I left it in the park overnight, and in the morning I left to go, uh… get some food somewhere… and when I came back the van was fuckin' gone! I called the cops and found out that *they* stole it!"

"Well, they didn't 'steal' it, Dion. They impounded it. Obviously, you had it parked somewhere where it wasn't supposed to be."

"Whatever, dude. I call it stealing. And now they want three hundred dollars before they'll give it back to me. You've got to help me out, man. If I don't get the money to them within the next three days they're gonna sell it with all my shit in it. That favor you did for me in San Diego, it'll all be for nothing if you let them take the van away from me."

I thought about it for a moment. Well, maybe he was right. I didn't want him stranded in Seattle, did I? "Listen, I've got to go somewhere right now. Call me back in a couple hours and I'll tell you if I can do it."

"Why can't you tell me *now*?"

"I've gotta go check my bank account. Is that okay?"

"Yeah… yeah, that's okay. But we've got to *hurry*. This is important."

"By the way, have you had any problems lately?"

"Problems? What do you mean?"

"With the fucking midgets. Problems with the fucking midgets."

"Oh. No, no. I haven't seen hide nor hair of 'em since Winona. Maybe I managed to drop out under their radar screen. Who the fuck knows?"

"Right," I said. "Anyway, call back in two hours."

"Okay, man, in two hours."

He hung up. I called Wanda and asked her what I should do. She told me not to do it. But, of course, I knew she was going to say that. I was back to where I'd started.

I checked my bank account and decided I could do without the three hundred even if I never got it back since the fall semester would be starting again within a few weeks. I had almost decided to go ahead and wire Dion the money when the phone rang again, about an hour before he was supposed to call back.

I picked up the phone.

It was Melanie.

Oh God, I thought. What *now*?

"Hey, when did you leave?" she said.

"I got a job offer to teach summer school, so I had to come back to California. Sorry. I was going to call you, but it was very sudden. You know."

"Well, that's great that you've got a job. You'd be proud of me. I've managed to stay away from Dion. Well, pretty much. I bumped into him about twenty minutes ago, but I only talked to him for a couple of minutes. I'm not getting into that van ever again."

"I know how you feel."

"I haven't done any more heroin since I saw you last."

"Ah. Well, that's fantastic. You see? You didn't need me after all. I knew you had the gumption. By the way, I just got done talking to Dion. He wants to borrow some money from me."

"Again? What for this time?"

I told her everything Dion told me.

Melanie cackled with amusement and said, "That's bull-shit. I just saw him sitting in his van. That's what he was doing in the park. Sitting in his van. He was cookin' Top Ramen in the back on that busted hot plate of his. Jesus, it smelled horri-ble. He was adding cayenne pepper to it."

"Wait a minute. *Twenty minutes ago* you saw him in the van? You sure it wasn't this morning, or—"

"I *literally* just left him. I'm tellin' you, he's the bullshit artists of all bullshit artists. He's a second-rate con man and always has been. Don't feel bad. At least you weren't fucking him for eight years."

"He's been fucking me way longer than that," I said.

About an hour later the phone rang again. It was Dion.

"So…" Dion said, wasting no time, "what about the money? Think you can give it to me?"

"Is the van still impounded?" I said.

"Yeah. Where the fuck else would it be? The clock is tick-ing, man." I could hear him snap his fingers. "Chop chop."

"Where was it about an hour ago?"

"I already *told* you. It's impounded. I don't know the exact address of the place."

"Why did Melanie just tell me on the phone that she saw you in the van only about an hour and twenty minutes ago?"

There was a long silence, then Dion said, "*What?* She must be mistaken, man. She must be thinkin' of the conversation we had *yesterday.*"

"No. It was today. She saw you eating Top Ramen with cayenne pepper in it."

"And that fuckin' bitch didn't even wanna taste it! I offered her some and she blew me off, man! She told me she was scared to get into the van! Can you believe it? What's to be scared of? Fuck that cunt! Can't you *see*? That bitch is just makin' up sto-ries to ruin my relationship with you—my best friend! When

I see her next I'm gonna fuckin' break that bitch's neck. This is the *last* time she lies about me, man! The *last* time!"

"Dion, have you gone completely fuckin' nuts? How could Melanie refuse to get into a van that wasn't there? What were you making that Top Ramen on, a rock with a lighter attached? How could she be lying about the van being there if you're saying it actually *was* there?"

"Wait a minute… now I'm all confused."

"I know you are. You're living in two alternate realities at the same time, Dion. Psychologists call it cognitive dissonance."

Dion began screaming into the phone. "That bitch is stealing my life! I'm gonna kill her! I'm gonna fuckin' kill her!"

I told Dion I had to go, then hung up.

I called Melanie on her cell to tell her that Dion was planning on killing her.

"Yeah?" she said. "What else is new?"

18.

I didn't hear from Dion again for almost a year. In the meantime I landed a full load of classes during the fall semester at CSU Long Beach. Everything went very smoothly from August to January without Dion in my life. I didn't think about invisible midgets or simian sharpshooters or using Boris Vallejo as a military weapon. A part of me had come to grips with the fact that my junkie friend from high school had been pulling my strings from the very beginning. I had caught him in one doozy of a lie, and could only conclude that everything that had preceded that lie had been equally fictional.

Now, of course, I knew *all* of it wasn't fictional. Lita Johnston existed. I had talked to her on the phone, after all. Why or when the lies started I really didn't know. I didn't care. I felt like

I was slowly getting over the post-traumatic stress syndrome instilled in me by Dion's paranoiac tall tales.

At the end of January, the Chair of the English Department at CSU Long Beach had to break the news to me that, once again, she didn't have any classes for me in the spring; however, she did me a favor by pulling a few strings and securing me a full load of Literature classes at El Camino Community College, where I had once been a student back in the early 1990s. It was in one of these classes that I met a student named Laurie.

Laurie was twenty-six, beautiful, and very talented in both writing and singing. We began chatting after class more and more, sometimes for a couple of hours. One day she just happened to tell me the following story, almost casually: During the summer of 2003 she had been making these collages, anti-Bush propaganda posters, that she had been hanging up all over Pacific Beach, the exact neighborhood where Dion experienced all his special brand of nonsense. She had been doing this, she said, for a few days when she suddenly noticed that people were following her around. She noticed it first, she said, when she walked into a 7–11 in Pacific Beach and realized that about fifteen people had followed her in and seemed to be tracking her every move. Some of them would even mimic her actions, as if to annoy her. She stopped putting up the flyers soon after that.

I asked her where this 7–11 was located.

She said, "The one on Garnet." The same 7–11 where Dion first noticed he was being stalked.

"This is blowing my mind," I said, and proceeded to tell her a brief version of Dion's story.

She didn't doubt a word of it.

Laurie soon revealed to me, via an autobiographical story she'd written, that she had suffered a schizophrenic breakdown when she was twenty-two. She was taking medication to treat her symptoms, which included paranoia. However, even taking Laurie's schizophrenia into account, what are the chances that I would meet the *one other paranoid person* in Pacific Beach who also hallucinated that a team of gang stalking jarheads followed her into the *same* 7–11 at around the *same* time after hanging up the *same* type of propaganda posters all around San Diego? The percentages boggled my mind.

Later that day, I had the urge to call Dion and tell him what I had learned, but of course I had no way of contacting him since I didn't even know where he was living now. So I just put it out of my mind.

The semester continued, and the after-class conversations with Laurie kept growing in duration. On the last day of class I asked her if she wanted to go out for dinner and she said yes. I knew she was scheduled to go on a trip to Ireland soon, so I said we could do it when she got back. But she wanted to get together as soon as possible. A day after the semester ended Laurie picked me up at my apartment and we spent the whole day together hiking the horse trails near her grandmother's house in Palos Verdes. We ended up spending the night together. The next morning she took a plane to Ireland and promised she'd call me the second she returned. She did, and pretty soon we were dating seriously.

That summer we spent together was more than idyllic. It was *Perfect* in every way. In August she confessed to me rather nervously that she wanted to marry me, something that other girlfriends had suggested in the past, but this was the first time I considered the possibility.

During that summer Dion wrote me a letter from Humboldt County, which was located about two hundred miles north of San Francisco. Somehow he had migrated down from Seattle to Humboldt, where his mother lived. I considered not responding, but the information that Laurie told me still gnawed at me. I felt he should know about it. I wrote him a short letter back telling him what I had been up to lately, and mentioned the story that Laurie had told me months before.

A few days later Dion called me on the phone. He sounded much better than he had in Seattle. He sounded a hell of a lot more calm. And I'm sure the Seattle PD were just happy to be rid of the Goldilocks Bandit.

Dion began the conversation by telling me that he hadn't had any recent experiences with the invisible midgets, but that he was actively pursuing more information about what had happened to him. He said he had been doing research on the Internet about breakthroughs in invisibility technology and was going to send me some printouts about it. I told him to do so, but wasn't particularly interested at this point. First of all, I wasn't sure I could even believe him, and second of all, I just wanted to put all the insanity behind me. I was more interested in spending time with Laurie than reading printouts from the Internet about invisible midgets.

Dion asked me to repeat the story that Laurie had told me in class, so I did.

"Jesus, how is that possible?" he said. "Maybe she was just making it up to impress you or something? I mean, did you say anything about me in class before this?"

"No, I didn't even mention your name. I never said anything about the invisible midgets or San Diego or anything."

"And she was living in Pacific Beach when I was living there?"

"Yeah."

"That's so weird. I knew a hell of a lot of people when I was there. I might know her. What's her full name?" I told him. "No, that doesn't ring a bell. Maybe I'd recognize her if I saw her. Hey, wait a minute." He pointed out to me that Laurie's last name was the Italian word for "police."

"Yeah, I think you're right," I said. "She made a joke about that."

Dion remained silent for a moment. "Don't you think that's a little odd?"

"What is?"

"It's like a big joke. Like they're flaunting it in our faces. Are you sure she's not some kind of plant?"

I sighed. "Yes, Dion, I'm sure of it."

"It's such a weird coincidence that I almost can't believe it. Maybe they're trying to figure out where the goggles are through you. Has she ever asked you about them?"

"Yes, she's asked me about them, but that's because I brought them up. They're the whole McGuffin of the story, fachrissakes."

"Did she seem unusually interested in finding out what happened to them?"

"Dion, I've known this girl for seven months now. I've been dating her for about three of those months. I think I would know if she was a fucking cop. Just drop it."

"Just be careful. That's all I'm saying."

I pushed the thought out of my mind and we moved on to a different topic. He told me how his mother was driving him up the wall, and that he was just living there until he could raise enough money to move out. He wasn't working, but he was collecting a check from the government each month. Apparently he had gone to a psychiatrist to help deal with the post-traumatic stress syndrome he incurred as a result of his run-in with the NCIS. The psychiatrist listened to the whole story, and immediately authorized Dion to receive a substantial crazy check from the government on the first of every month. Dion was making more money than *I* was teaching a full load of classes. He often referred to himself as "retired."

"Well, at least you got something out of it," I said.

"Fuck, man, I'm just takin' back what's *mine*! I'm not guilty. You know how much time and money they made me lose with all their fuckin' bullshit? No amount of cash can ever get Jessica back. Jessica left me because of *them*!"

I failed to point out that he had kicked Jessica out of his apartment long before the incident with the night vision goggles had ever even occurred. We talked for a little longer, then I had to leave because I had a date with Laurie that night.

All throughout the date my mind kept drifting toward the possibility that Laurie was a cop who had been assigned to spy on me. At one point, while we were eating dinner at The Green Temple in Redondo Beach, she nonchalantly asked a question about Dion and I snapped, "Why do you want to know that?"

She looked confused. "Because… you mentioned that you talked to him earlier."

"Oh. Right. Sorry. I'm just on edge. Dion does that to me." I told her everything Dion said, except the part about her being a spy. I didn't want to offend her. Besides, it was so ridiculous I didn't even want to mention it.

Laurie eventually moved in with me, and we lived together for about five months. We loved being around each other, and the entire situation would have been *Perfect* except for the fact that she had a schizophrenic breakdown near the end of August and ended up locked inside a psych ward for several weeks.

I was devoted to Laurie so much I visited her every day in the hospital with the hope that she would recover and be the same laughing little sprite I had so deeply fallen in love with.

One symptom of her breakdown was extreme paranoia. She began accusing me of absurd crimes: that I was abusive, an international terrorist, a Satanist, and the man who had manipulated Timothy McVeigh into bringing down the Oklahoma City building. The vast majority of those accusations were untrue. I seemed to be trapped into some kind of weird karmic cycle, but I wasn't quite sure what I had done to deserve all this. More than once my mind wandered back to an obscure Fritz Leiber story called "Schizo Jimmie." It's about a guy named Jimmie who's sort of like Typhoid Mary. Typhoid Mary was capable of spreading typhoid without exhibiting any symptoms herself. Schizo Jimmie, on the other hand, was capable of spreading *schizophrenia* without exhibiting any symptoms himself. Was *I* Schizo Jimmie? It seemed like everyone else around me was crazy, or eventually became so even if they weren't when I first met them. When I first met Laurie I was so happy to have found a normal girlfriend at last. That normalcy only lasted about three months before being destroyed by the onset of reality. I had known that Laurie suffered from schizophrenia in the past, but she seemed so balanced now I assumed it was all behind her. That was just wish fulfillment on my part. Either way, my *Perfect* little dream world came crashing in on me right before the beginning of the fall semester. I had to get ready for my new classes while also visiting my girlfriend in the psych ward. In the midst of all that, I was also moving to a new apartment in Long Beach in order to be closer to campus.

I had now begun speaking to Dion regularly and told him everything that was going on and how stressed out I was. Dion offered to drive down from Humboldt in his van and help me move all my shit. He was the only one I knew with a van that could do the job—and besides, it was half my van anyway—so I said okay.

He arrived on a Saturday and the move went smoothly. After being released from prison back in '99, for a few months Dion had gotten a job as a professional mover in Minnesota, so thanks to him we were able to maneuver furniture up three flights of stairs and through a narrow doorway far easier than I would've been able to do with anyone else. Concentrating on the move made not knowing where Laurie was much easier. The hospital had let her outside to smoke and somehow she had broken free. No one knew where the hell she was now.

Dion slept on the carpet during my first night in the apartment. I was grateful for his helping me move, but to be honest I was frightened he would never leave. I knew the Dion curse well. And I suspect he might've settled in for permanent residency if not for the fact that Laurie showed up on my doorstep early the next morning. This was the first time the two of them met. Laurie was so whacked out of her mind that Dion was scared shitless of her. I had to leave the two of them alone at one point in order to get something out of Dion's van. Whatever she said to him must have spooked him because he seemed anxious to take off the second I got back. I tried to convince Laurie to return to the hospital but she would have none of it. I made her a sandwich, which she ate on my floor like a monkey with a banana it didn't want ripped away. She had completely devolved from the vision I had fallen in love with so many months before.

She asked to borrow twenty bucks so she could get a bus ride somewhere, but she wouldn't tell me where. Despite the fact that I begged her to stay with me she wouldn't listen. She would alternate between accusing me of signaling to sharpshooters stationed on the nearby rooftops to being all lovey-dovey with me. It was very confusing. At one point she accused me of carrying on some kind of illicit homosexual relationship with Dion. I didn't understand where this was coming from. The only times I was able to break through her madness was when I was able to use humor on her. I would take something she said, some paranoid accusation, and instead of denying it I would magnify it a thousand fold.

"Yes, Laurie," I said to her at one point while strolling with her along Ocean Boulevard (only a few blocks away from my new apartment), "I *am* a Satanist. I really did blow up the Oklahoma City building. Oh, and guess what? I'm also a werewolf. Did you know me and my S.S. Werewolf Attack Battalion were responsible for the disappearance of Chandra Levy?" During these moments of absurdity she would smile, laugh, and regain her sanity. But only for a moment. I thought it was interesting that sarcasm and absurdist humor were the only tools capable of breaking through her schizophrenia. The fact that I succeeded in breaking through for even a moment is amazing.

I hugged Laurie in downtown Long Beach, gave her the twenty, and let her walk away to wherever the hell she was going. I knew I couldn't hold her. Any attempt to do so would just make her even more distrustful and paranoid and, yes, violent.

I went back to my apartment and began putting all my shit away, wondering how the hell our love affair had fallen apart so drastically. Perhaps most frustrating was the fact that it was no one's fault. We had been *Perfectly* happy. Then *boom*: Laurie's brain turns inside out. It was unfair of the universe to do this to me. I felt cheated.

Dion called a little while later and asked me if Laurie was gone. He said he'd met some crazy people in his life, but Laurie had freaked the shit out of him. He was genuinely scared of her and the spacey look in her eyes. He said he was shocked I was able to stand it at all. Of course, I was able to stand it because I knew that wasn't the way she really was.

Dion tried to convince me to forget about her. "Relationships are hard enough without throwing schizophrenia into the pot. You're not going to be able to cure her of this. It's going to keep coming back and back and back. It's going to be hell. You need to cut your losses." I knew that was the sensible thing to do, but I just couldn't bring myself to do it.

I was in love with her. Simple as that.

This stressful situation was not helped by Dion always suggesting that dark forces in the U.S. government had sent Laurie into my life on purpose in order to disrupt it. Why they would do this I had no idea, but Dion seemed to believe they hadn't given up looking for the goggles.

While we were sitting on the floor of my apartment eating take-out food from Super Mex and ignoring the horrible Ben Affleck movie (*Paycheck*) playing in the background, I asked him if he'd experienced any harassment since leaving Humboldt.

"Something strange happened to me when I stopped off in San Francisco," he said. "I was driving down Geary listening to 'Pet Sematary' by The Ramones when a van that looked just like mine pulled up next to me at a stoplight. You know what was playing in the van? 'Pet Sematary' by The Ramones. The driver just looked at me and smiled. Now, what are the chances of that? I think it was a message. They were just letting me know that they were still on my ass."

At times like these I wondered if maybe Dion didn't deserve that monthly crazy check. My honest opinion at this time was that everything Dion had told me about his San Diego experiences was true (more or less). But, of course, a situation like that would *drive* you into paranoia. How could it not? I suspected Dion was now suffering from clinical paranoia brought on by the fact that a large group of strangers really had been persecuting him at one point.

When I got home from work later that day I called both of Laurie's parents, as well as her friends, but none of them had heard from Laurie. I was the last person who had seen her. I spent hours calling every psych ward in the area. At last, I found her in the psych ward at Harbor General in Torrance. I talked to her only for a second. She sounded messed up, but she did recognize my voice and asked me to visit her as soon as possible. I promised her I would, then called her parents and told them I had located her at last.

I think partly because of the visit from Laurie, and partly because of the fact that I was going to make him do more work for me if he decided to stay, Dion took off in his van the very next morning right before I had to go to work. (It was difficult paying attention to my classes with all this chaos going on, but somehow I managed to do it. I never even missed a day.) He had some friends he wanted to visit in Southern California before he returned to his mother's house in Humboldt. He said he'd be back to see me again soon. Just before he left, he told me to be careful and keep my eyes open. He said that the gang stalking activity often seemed to follow in his wake, so he wanted me to tell him if anything strange or out of the ordinary should happen to me while he was in town.

Only a few hours after Dion took off I received a call from Ziva Charnis. Ziva had been a student of mine in the very first English class I had ever taught back in the fall of 2002. She had moved back to New York in the spring of 2004, but we had kept in touch ever since. She wanted to know if we could go out to lunch together while she was in town visiting some friends from CSULB. Plus, she wanted to see my new apartment. I said that would be fine. I was supposed to visit Laurie in the psych ward early in the morning, so I figured I could rendezvous with Ziva in the afternoon.

At around 3:00 P.M. Ziva called to tell me she was taking the Passport D bus from the campus to my apartment. She said she'd call me when she reached the corner of Alamitos and Ocean. I told her I'd pay for lunch.

I was very much looking forward to having a relaxing lunch with Ziva, just the two of us. I thought of her as some kind of refuge from the madness swirling around me.

About twenty minutes later she called to tell me that she was standing on the corner of Alamitos and Ocean with a man she'd just met on the bus. The man was named Pete and he had offered to buy lunch for the both of us.

"What?" I said. "Who the fuck is *Pete*?"

"He's a longshoreman," she said. "He seems like a nice guy."

"What the *fuck*? Are you kidding? I don't know this guy!"

"Oh, come on, it'll be fun." Ziva had always been impetuous in that way, so naively sociable that it disturbed me. I had always been the exact opposite, even before I met either Dion *or* Laurie.

Reluctantly, I agreed to meet the both of them on the corner. There was no way out of it. I was pissed at Ziva for putting me in this awkward situation, but I wasn't thinking about anything relating to what Dion had told me. Nothing at all.

I saw the two of them standing on the corner of Alamitos and Ocean. Ziva looked like she always looked, a cute little, dark-haired waif with striking Middle Eastern features (she was from Israel). Pete was in his late thirties or early forties, wore nondescript clothing, had dark hair, appeared to be Mexican but could have been Salvadoran or something similar (I'm not sure about that). He seemed to be somewhat overweight. When I reached out to shake his hand the first thing I noticed was that he appeared to have a manicure. His fingernails were pristine and his palms were as soft as a baby's. "So you work at the docks?" I said.

"Yeah," he said. "Nice to meet you."

"You live in the area?"

"Right around the corner," he said.

"Yeah? And you just bumped into Ziva on the bus?" I asked.

"Oh, I ride the Passport D all the time."

"So do I. I've never seen you."

"Well, I'm usually on it at night."

I turned to Ziva. "So what's the plan?"

"Pete's going to buy us lunch at 555." 555 was a very expensive restaurant on the corner of Ocean and Linden.

"Why are you going to do that?" I asked.

"Oh, man, I'm goin' through a divorce now and I'm just trying to be sociable again."

"Yeah?"

Everything he said seemed insincere to me. It was at that point that my mind retrieved Dion's warning. I was trying very hard not to be paranoid, but it became increasingly difficult.

We walked to the front door of 555, where you need a reservation, but this schlub told us he only had to go talk to the manager and we'd be given a table. Why that would be I had no idea. None of us were dressed for 555. I had been hoping to go to a simple coffee shop with Ziva, or a café, not a candlelit dinner for three. The second Pete left I turned to Ziva and said, "What the fuck are you doin'? Are you tryin' to *kill* me?"

"What're you talking about?"

"Did you approach him on the bus or did he approach you?"

"He approached me."

"Yeah? Just like that? I ride the Passport D all the time and I've never seen his face."

"So? You probably weren't paying attention. What're you saying, that's he's lying about riding the bus? Obviously he rides the bus. I met him there."

"I understand that, but... I don't know... there's something weird about all this. Don't you think there's something weird about all this?"

"What's weird about it?"

I kept telling myself not to be paranoid. Ziva barely knew what was going on with Laurie. I had told her about the invisible midget scenario once before, but she seemed to think it was nuts. I couldn't figure out how to explain why this guy was making me nervous. It was just a feeling.

Pete came back from his conference with the manager (whose name, oddly enough, was "Fink") and told us our table was secure. He said his brother did business with 555 all the time, so all he had to do was mention his brother's name in order to get a table. I asked him what his brother did, but he was short on specifics. As all three of us stood there waiting he kept trying to spark up a dialogue with me, almost ignoring Ziva, which is odd if his initial purpose in approaching Ziva on the bus was to get laid. As my comments to Pete grew more sarcastic, Ziva began to grow more and more uncomfortable and was no doubt wishing she hadn't started this mess in the first place. At one point Ziva told us she had to go call someone real quick and walked around the corner of the building, leaving me alone with this guy *she* had picked up off the bus.

Now here we were staring at each other. Suddenly, Pete told me he had been doing a lot of reading since his divorce.

"Yeah?" I said. "What've you been reading?"

"I just came across this book called *Behold a Pale Horse*," he said. "Have you ever heard of it?"

I couldn't believe it. The odds of this random stranger mentioning that particular book at that particular moment boggled my mind. On top of everything else, this just seemed to be the nail in the coffin. From that millisecond on I was convinced he was a government agent.

"*Behold a Pale Horse*," I repeated with no inflection in my voice. "By who?"

"God…" He snapped his fingers. "I can't remember his name."

"You can't remember his name. Is it a novel or non-fiction?"

"Oh, it's non-fiction. It's about conspiracies and shit. Milton William Cooper. Yeah, that's it. That's his name."

"Yeah." I laughed. "William Cooper. And you just got done reading this book, did you?"

"Oh, yeah, man. I think all that shit's true. He talks about secret societies taking over the world and shit. You know anything about that?"

I just stared at him. I almost said, "Yes, I'm a 32nd Degree Scottish Rite Freemason. And you don't have to worry about it, because we've already taken over the world."

"No," I said, "not really. Why don't you tell me about it?"

He seemed flustered now, as if the script hadn't allowed for this possibility. "Oh… well, you know, he talks about how aliens are abducting people and shit. Performing experiments on them. You know. You've read any books about that shit?"

I had whole bookshelves devoted to that one subject. "No. Can't say that I have, but I'm interested. Tell me more about it."

He just gave me generalities. Occasionally he'd hit upon specifics: "The aliens and the secret societies, they're workin' together. They had a meeting beneath the glaciers in the Antarctic."

I guess that was the one paragraph he'd read in Cooper's book. "Wow," I said, "you should tell that one to Ziva when she gets back."

"I will."

Ziva arrived on the scene at the same moment as the maitre d'. He referred to Pete by his first name, then led us to a table in the middle of the room. The waitress gave me a menu, but I wasn't looking at it.

"Tell Ziva what you just told me," I said.

"Well," he said, "there's this book called *Behold a Pale Horse* by Milton Cooper. Have you ever read it?"

"No," Ziva said, "what is it?"

I vividly remembered showing Ziva a copy of the book when she was in my bedroom one time, but apparently this had not made any particular impression on her.

"It's this book about how certain secret societies—you know, the Illuminati—are working with the space aliens to enslave the human race."

"That's bullshit," Ziva said in her inimitable manner. She was raised in New York and wore the attitude on her sleeve.

Pete seemed nonplussed. Perhaps they hadn't prepared him for this. "Well… you know, maybe." He stuck to the script anyway. What else was there to do? "You see, the Illuminati—"

"Isn't that a code word for the Jews?" Ziva said. "You know I'm Jewish."

"Oh, are you?" Pete said. "I didn't know that."

"Does that make a difference?" Ziva said. "I know *I'm* not in charge of the world, and my parents were born in Israel."

"Well," Pete said, "when Cooper uses the word 'Illuminati' I don't think he's referring to only the Jews. In fact, he says if you read *The Protocols of the Elders of Zion*—"

"Wasn't that a hoax? Can I have some white wine?"

"Yes and no."

"Yes she can have some white wine and no she can't?" I said.

"Oh, no," he said, "yeah, you can have some wine." He paused for a moment, then looked at me and said, "Do you want anything?"

"I don't drink," I said.

This seemed to throw him. Perhaps they got that wrong in the file—again. "Iced tea?" he said.

"I don't drink anything," I said.

Pete launched back into *The Protocols of the Elders of Zion*: "You see, if you read *The Protocols* and just cross out the word 'Jew' and put in the word 'Illuminati' then you'll see that most of it is true."

"So it's a code word for 'Jew,'" Ziva said. "I was right the first time."

"No, no," Pete said. "What I mean is, have you ever heard of the Freemasons?"

Ziva laughed and gestured toward me with her thumb. "He's one."

"Oh, really?" He didn't seem surprised by this. "That's interesting," he said. "I'd like to know more about that. What is it exactly that you people do?"

"As far I can tell," Ziva said, "they sit around and eat potato salad all the time."

Pete laughed. The first genuine laugh of the evening. "Well," he said, "I don't think Milton's saying all Freemasons are involved in this stuff. Just some of them. Don't get me wrong. I'm not totally against secret societies. The way this government is going, we may have to form our own secret society in order to fight back against the oppression. We need anarchy in this country if it's ever going to get any better."

"Everything seems fine to me," Ziva said. "I'm not starving. Speaking of starving…"

At that point the waiter came by and Pete took it upon himself to order for both of us. I already knew I wasn't going to touch any of the food or water that was served to me here. I was seriously considering getting up and leaving Ziva alone with him, but I just couldn't bring myself to do it.

For the next half hour we continued to engage him in this bland conversation that went nowhere. Anyone who'd really read *Behold a Pale Horse* and was interested in the contents would be able to come up with some unique or unusual insights. This guy seemed to be quoting the liner notes from *Conspiracies for Dummies*. He kept trying to draw me into a conversation about the Bush administration and how out of control it was. I told him I voted for Barry Goldwater in the last election. He didn't know what to make of that. Ziva was growing confused and annoyed by my overt hostility toward this man.

The lunch ended awkwardly when Ziva decided to terminate it by saying that she and her friend (me) had to leave to take care of some business. The dude seemed real disappointed, but I wasn't sure if it was because his mission had not gone as planned or because he was a desperate human being trapped in a world of pain, trying and failing to connect with someone but being foiled by some paranoid guy he'd never met before. I still haven't figured it out.

The second we hit the sidewalk Ziva turned to me and said, "What the fuck's wrong with you?"

"What?"

"What was all that about? He was trying to do us a favor."

"Are you kidding? You don't know what's been going on in my life. He must've been tipped off that you were on the way to my house, intercepted you, then tried to pull some kind of psychological warfare game with me. You didn't hear what he was saying when you left me alone with the guy—oh, thanks for that, by the way. The second you leave, the dude brings up *Behold a Pale Horse*. What the fuck's that? Supposedly he rides the Passport D all the time. I've never seen him on that bus. The guy's a plant. How else do you explain it?"

"Uh, I explain it like this: You've been hanging around with so many crazy fucking people that it's rubbed off on you. You need to get away from those maniacs. They're drivin' you over the edge."

For someone who was so damn small she sure had a mouth on her. For a second I felt horrible. Maybe I really had ruined the guy's day. But then all I had to do was retrieve the memory of the fumbled *Behold a Pale Horse* nonsense and all the bullshit about "we need anarchy in this country" to convince me I was right.

"I don't know, Ziva," I said, "you don't know everything that's been going on."

"I know all about the invisible midgets. What else is there to know? Let's just go to your apartment and drop it, okay?"

So we did just that. I showed her my apartment, we hung out for a little while, I told her a little bit about the whole Laurie situation, then she left. Our lunch hadn't quite turned out the way I had planned.

By the way, a point of interest: I lived in that apartment for two years, from September of 2005 to October of 2007. I then moved only about a mile away. I still take the Passport D to work. Never once have I seen "Pete" again, despite the fact that he supposedly lived right around the corner from my building, despite the fact that he supposedly rode that exact bus every day. I wonder how his longshoreman gig is working out for him these days.

19.

I never told Dion about my encounter with Pete for fear that it might send him spiraling down into paranoid-fueled hysteria. He returned to my apartment a day later, helped me move some more stuff, then took off back for Humboldt, but not before warning me to be careful around Laurie. He suggested I be there for her as a friend, but not resume my relationship with her. Other friends told me the same thing. I listened to the advice and promptly ignored it. As they knew I would. I'm pretty stubborn at times.

The Laurie situation continued to spiral out of control. When she was released from the psych ward for the last time (at least the last time during the period that I knew her) she asked to move into my new apartment with me. I let her. Her doctor told me to hide all the knives.

We stayed together for five more months. Our relationship continued to sink into utter chaos. If Laurie didn't show occasional signs of recovering, of returning to the personality I had fallen in love with, I probably would've lost all hope. But I continued to hold on to that fragile, silken strand against all logic and reason.

The relationship ended. I didn't end it. I never would have, no matter how bad it got, no matter unsafe and unpredictable the situation became. For some reason I felt responsible for her insanity, despite the fact that her madness began long before she ever met me. Why I put up with all that chaos I have no idea, except for the obvious fact that I loved her. It didn't matter that I almost lost my life because of her. Several times, for example, she almost killed us in her car, and yet I insisted on climbing right back into that damn passenger seat.

Dion stayed away from Long Beach this entire time. There was no reason to ask why. The second she took off for good, in February of 2006, he asked me if he could drive down in his van and visit for a couple of days. I said sure, mainly because I didn't know how to say no. I did need to get my mind off the situation. I felt very much betrayed. I had done so much to help Laurie, and all I had received in return was mistrust and bitter recriminations. But I was also scared that I was replacing one wild card with another.

When Dion called I was putting the finishing touches on a novel I had begun while dating Laurie. Months earlier the novel had been interrupted by her committal to the psych ward. Her abrupt departure prompted me to pull out the manuscript and finish what I had started. Though I was only halfway done, I finished the entire novel in a very short space of time. The novel was called *The Goblin Hordes of Pacific Beach*. I started the book while Laurie and I were still in our idyllic stage, before the madness encroached. It began as a fictionalized version of the night vision goggle incident, then quickly became a commentary on the thin line between sanity and madness with the introduction of a character based partly on Laurie (and partly on another eccentric girlfriend from my recent past).

I put the finishing touches on *The Goblin Hordes of Pacific Beach*, just a few hours before Dion arrived, and began sending off queries to potential publishers. The first draft of the novel was dedicated to Laurie. After Laurie left I said "Fuck it" and changed the dedication to "To Invisible Midgets… Everywhere."

I received a response from one agent within twenty-four hours of sending out the query. She asked to read the first three chapters. I sent them, but she eventually responded that she couldn't represent the novel because she made it a rule to only represent novels in which the main character was someone she herself would be interested in *dating*.

When Dion pulled into my driveway he was all excited about something. His research into recent breakthroughs in invisibility technology had never ended. He had now compiled a whole file on this topic. He urged me to come upstairs with him so he could show me a website he had stumbled across.

It was the website of a man named Richard Schowengerdt, someone I'd never heard of. Schowengerdt claimed to be a scientist who worked on Top Secret projects at Northrup Grumman. Aside from his work at Northrup he had a personal project: the development of invisibility technology. He had even filed a patent on the subject. Anyone just coming across the site by accident might assume he was a nut. After all, if you'd really developed such a thing why the heck would you advertise the fact on some dinky little website? I myself wasn't quite sure what to make of it; however, what caught my attention was all the references to Scottish Rite Freemasonry on the site. It seemed like a complete non sequitur, and yet there it was: a classic crank mishmash of conspiratorial hot topics. I noticed that Schowengerdt lived in Costa Mesa, and claimed he was a 33rd Degree Scottish Rite Freemason. I knew the closest Scottish Rite Lodge would be the one in Long Beach on Elm Avenue, located within walking distance of my apartment building. If this gentleman attended the Lodge regularly, that meant I'd no doubt met him. And yet I couldn't recall ever having done so.

Dion suggested I use my "Masonic mumbo-jumbo" (his words) to contact the guy and see if he knew anything about what had happened to Dion in San Diego. It seemed like a stretch, but what the hell. I was game. Why not? It was a good way to get my mind off the whole phantasmagoric Laurie affair that was still weighing on my mind.

Not expecting a reply, on March 7th, 2006, I sent
Schowengerdt the following email:

Dear Mr. Schowengerdt:

*A colleague recently drew my attention to your website.
I'm very interested in interviewing you about your inventions.
Previously, I've done interviews with a wide range of experts
in their particular fields, including David Ulin, the current
editor of the* L.A. Times Book Review; *Dr. Stephan A.
Hoeller, Bishop of the Gnostic Church in Los Angeles; Paul
Laffoley, one of the most accomplished experimental painters
working today; and Robert Dobbs, the former archivist of
media scholar Marshall McLuhan.*

*My short stories, articles, and interviews have appeared
in such magazines and anthologies as* After Shocks, The Chi-
ron Review, Like Water Burning, Mysteries, New Dawn,
The New York Review of Science Fiction, Paranoia, The
Pedestal, Riprap, Steamshovel Press, *and* The Third Alter-
native. *I'm currently teaching English Composition, Literature, and Creative Writing at California State University at
Long Beach. Since I know your office is not too far away, I
thought I would ask if my photographer and I could drop by to
record a brief interview with you.*

*Besides living very near one another, I couldn't help but
notice that we also have one other detail in common: While
you're a member of the Newport Beach Masonic Lodge, I'm a
member of Torrance University Lodge #394. (I'm also a member of the Valley of Long Beach.) If you're willing, perhaps this
interview could also encompass esoteric matters such as how
Masonry has influenced your life and work. Since I'm a member of the Scottish Rite Research Society, I'm always interested
in documenting the significance of Masonic thought in society.
For example, Bishop Hoeller, a 32nd Degree Freemason like*

myself, talked at length about how Masonry has influenced his scholarly career.

If you're willing, please let me know when you might be available for such an interview. I look forward to hearing from you.

Sincerely yours,
Robert Guffey

A day later I received the following reply:

Dear Mr. Guffey,

I would be pleased to meet with you for an interview concerning Project Chameleo. During the past year or two there has been increasing interest in the concept, mostly among the military but also occasionally from the commercial advertising sector. As microprocessor and display technologies continue to advance Project Chameleo becomes increasingly feasible.

It is of particular significance to me that you have interviewed such persons as Dr. Hoeller, a friend and teacher of mine, and that you are a Scottish Rite Mason. I have attended many of Dr. Hoeller's lectures at the Philosophical Research Society and greatly admire his wisdom, wit, and knowledge of the Kabbalah and other esoteric subjects. However, I really didn't know that he was a Mason as I have not had very many personal conversations with him. I visited his Church one time over twenty years ago but there were other persons present so I have not had an opportunity to develop a close friendship with him. Certainly, I think it would be appropriate that we encompass such subjects as Masonry and other esoteric streams of thought in our society today. I was a Rosicrucian (San Jose Clan) for over 25 years and have also studied the Kabbalah, Cabala, or Qaballah as some prefer to spell it under different schools such as the Builders of the *Adytum*. In addition, I have published a few articles in The New Age, Long Beach Scottish Rite *newspaper, and other fraternal publications on esoteric subjects. I am currently still active in degree work at LBSR (12th, 14th, 15th, 17th, 18th, 19th, and 32nd degrees). I was formerly active in others but some degrees have been discontinued or seldom presented due to general decline of membership. I am also an avid believer in the extraterrestrial presence and agenda in our society but if you think this is too controversial we could leave that out.*

Concerning the place for the interview, it would perhaps be best if we did it at the University as I do not have a good place to meet at our contractor-provided facility in El Segundo. I work for the Federal government but we are provided with only limited spaces at Northrop Grumman El Segundo and as it is under tight security control I could not bring unauthorized visitors into the facility. We could, of course, meet in a restaurant but may not have the privacy and quiet surroundings necessary for recording. I pass by the University every day on the way to work as I live in Costa Mesa so that would be convenient for me. Late afternoon or evening would be best for me, sometime on Mondays through Thursdays. I have a three day holiday every weekend so we could also do it on a Friday or Saturday during the day if that is better for you.

Just provide me with a few alternate dates and we can work out a mutually agreeable time and place. I am providing some alternate phone numbers for you below. Thank you for your interest in Project Chameleo.

Best regards,
Richard N. Schowengerdt

The opaque die was cast. I wasn't certain if Dion and I were going to be meeting with a nutjob or not, but the fact that he was a 33rd Degree Freemason in my own lodge seemed to suggest to me that he was far from unstable. Also, I began digging up any information I could find on Schowengerdt and discovered this piece from *Defense Review*, a very well-respected publication among military, intelligence, and law enforcement officers. In the 3-3-05 edition the CEO of *Defense Review*, David Crane, wrote the following in an article entitled "Robo-Soldier Ready for Combat Deployment to Iraq for Urban Warfare/CI Ops":

Hopefully, the U.S. Army and U.S. Marine Corps will soon have operational and effective electro-optical camouflage, a.k.a. adaptive camouflage, a.k.a. chameleonic camouflage, a.k.a. cloaking technology for infantry soldiers/Marines and combat systems. It's our [*Defense Review's*] understanding that the Army is currently working on it. If we were them, we would be looking very closely at Project Chameleo, which is run by Richard Schowengerdt, along with the adaptive camouflage sensor-and-display systems tech being worked on by Philip Moynihan of Caltech and Maurice Langevin of Tracer Round Associates, Ltd. for NASA's Jet Propulsion Laboratory. Both projects are essentially "Predator" camouflage brought to life.

The fact that a journal like *Defense Review* would so vigorously support Schowengerdt's research indicated that he was by no means a nutjob. Nonetheless, the likelihood that his research would in any way overlap with Dion's experiences in San Diego seemed very low indeed.

I wanted someone other than Dion to take photos during the interview, as the idea was to publish the interview in a professional magazine, just as I'd done in the past with other interviews. I knew there was a college student who lived downstairs whose major was photography. Her name was Melissa Oliver. I had been talking to her in the halls over the course of several months and seemed to think she was interested in me, but I could never follow through on this hunch because I was dating Laurie at the time. Now that Laurie had divested herself from my life, I felt free to pursue the matter further. So, asking her to come along to take photographs actually served two purposes at once. I told her the exact subject matter of the interview. She seemed to be neither skeptical nor credulous. In fact, she seemed to be up for almost any crazy possibility. This just made me all the more interested in her.

With my Mission Impossible team assembled, I agreed to meet Richard at the Scottish Rite on 855 Elm Avenue in Long Beach. Richard was performing in one of the degrees, the 32nd (the Master of the Royal Secret degree), and needed to be there until about noon. The fact that Richard was performing the degree made me even more curious. I'd seen the 32nd degree enacted several times, and the Lodge generally used the same performers over and over again. I *must* have seen this man perform the ritual, but didn't remember him at all.

When in the dining hall of the Scottish Rite at around 7:30 A.M., I glanced around the room and saw some people I'd met in the past. For a period of three weeks back in the spring of 2005 I had performed intensive research in the Research Library thanks to the generosity of the head librarian, John McKenzie. I now saw John sitting with a large group of older men near the front of the hall. John spotted me and waved me over.

"Hello, Robert!" he said. "Are you performing in one of the degrees today?"

"No," I said. "I'm mainly here to meet Richard Schowengerdt. I'm interviewing him. Have you seen him around?"

He looked puzzled for a moment, then said, "You know, I thought he was just here. Where did he go?" No one at the table knew he was gone. No one had even seen him leave. I began to wonder just how far this man had progressed in his invisibility research.

John asked me why I was going to interview Richard.

"I'm going to ask him questions about his research into invisibility technology."

Once again, he seemed puzzled. "What?"

I was surprised that he wasn't aware of this. "Richard's never told you about that?"

John sipped at his coffee. "That never came up," he said.

"How long have you known Richard?"

"Oh, about thirty years now." No one else at the table had ever heard about Richard's research either. You'd think that would be a colorful enough topic to bring up in casual conversation at some point or another.

Eventually I learned that Richard had gone backstage to change into his ceremonial robes. So I excused myself from the table and made my way up a narrow flight of stairs to the changing room where all the actors ready themselves for the performance. In the changing room I asked someone if they could point out Richard Schowengerdt for me. They pointed at a man about five feet away from me dressed exactly like the Pope. He was adjusting his mitre in the mirror when I approached.

"Richard?"

He turned around. "Ah, you must be Robert?"

Richard struck me as an affable, grandfatherly gentleman. His rural roots were belied by his bolo tie and his polite, Southern manners. He had a full head of graying hair, a boyish smile, and a dry wit. Despite the fact that he was in his seventies he seemed enthusiastic, full of energy, eager to talk about his ongoing project to make men invisible. He appeared to be a quiet, modest gentleman who was wise enough not to boast too much about how much he really knew about the esoteric secrets of the world.

I realized that I had indeed seen this man before. I'd seen him perform the 32nd degree many, many times. I distinctly remembered wondering who he was the first time I saw him perform, as his voice was deep—with a slight hint of a Southern drawl—and his tone authoritative. His delivery projected a certain authenticity, which was rare, as the degrees were often performed on the fly these days. The members who were willing to devote the necessary time to perform the rituals properly were dying off, and no younger Masons seemed willing to step forward and fill the void. Many of the actors were roped in at the last minute and had to read their lines off cue cards. Richard was one of the few who had his lines memorized.

Richard shook my hand with a tight grip, told me he was glad to meet me and looked forward to talking to me at length right after the ritual. He suggested we go out for lunch first, then conduct the interview afterward. I said that was fine, and that my friend Dion would be coming along as well.

A man came into the room and demanded that all the actors file out onto stage. They were going to do a run through before the start of the actual ritual. The ritual was conducted pretty much the same way I'd seen it many times before, and as usual Richard's performance was the most refined.

After all was said and done and the virgins duly sacrificed, Richard and I left the Lodge (before the ritual was even over, in fact) so we would have time to have lunch before the interview. Richard led me out to the parking lot where I saw his car for the first time. I immediately noticed the license plate: CLOAKER1. I laughed and asked him if the car itself turned invisible. He said he had plans to develop such a thing, but it would require a lot of money to manufacture the product itself. He then mentioned that the makers of a recent James Bond movie had based their invisible car on his concepts. I didn't know whether to believe this or not, but at that moment I saw Dion's black death van pulling up outside the Lodge. I had asked him to meet me there at noon, and for once he was on time.

I pointed at the van and said, "That's my friend Dion." Richard walked over to the van, shook Dion's hand, and told him he was very pleased to meet him. I could tell that Dion was a little nervous. He had confessed to me the previous night that the prospect of meeting Richard and having some of his nightmares confirmed was bringing forth extremely bad memories and that he was now experiencing what could only be described as outright panic attacks as a result. (For those of you who still think he was making up all this shit, consider the fact that people don't suffer from panic attacks as a result of experiences that they *themselves* have willfully conjured out of their own imaginations—that is, unless Dion made it all up without knowing he was doing so, but I think that's a stretch. As Sherlock Holmes said, "Once you eliminate the impossible, whatever remains, no matter how improbable, must be the truth.")

I asked Richard if he'd like to eat at George's Greek Café on Pine Avenue, one of my favorite restaurants in Long Beach. He said he loved Greek food so that wouldn't be a problem. Richard, Dion, and I proceeded to have a pleasant lunch. He was a very likeable guy, down-to-earth and gentle, but also very intelligent. No doubt about it. I knew now I wasn't going to be interviewing a nut. I still wasn't sure that any of this was going to tie in with Dion's experiences, but I knew I would get a good interview out of it at least. During lunch, Dion and I were careful not to mention invisible midgets or anything even vaguely related to what had happened in San Diego. In fact, I warned Dion not to do so. I just wanted to conduct the interview and see where it took us without any influence on our part.

After lunch Dion and I drove to the CSU Long Beach campus with Richard trailing close behind in his Invisicar. I had decided the best place to conduct the interview was in my office on campus. Since it was a Saturday I knew we wouldn't be interrupted. I had asked Melissa to meet us there. When we arrived she was sitting outside my office door with her tripod and camera. I introduced Richard to Melissa, then we settled down for the interview.

* * *

INTERVIEW WITH RICHARD SCHOWENGERDT
(3-11-06)

G: ROBERT GUFFEY
S: RICHARD SCHOWENGERDT
DF: DION FULLER
MO: MELISSA OLIVER

G: Can you just briefly give us a rundown of your professional background? I know it's a lot.

S: Are we on the tape?

G: Yes.

S: Okay. I'm an engineer by profession. I also have a bachelor's degree in business administration, and I've been in the industry, the aerospace industry, for over forty years and in and out of the government, so I have a lot of diverse experience as an electronics engineer.

G: I guess the main question is, what is Project Chameleo?

S: Project Chameleo was conceived with the idea that we could take the background and present it on a surface surrounding an object or in front of an object, and portray the background on this image, effectively giving you the same effect as if you were looking through the object, seeing the background behind it. And the chameleon does this. That's why I chose Chameleo, which is the Latin word for the genus chameleo, or chameleon.

G: When did you first come up with the idea for all this?

S: I came up with the basic idea in 1987, '88, along in that time frame, and I started working on the idea for a patent. And I finally secured the patent in '94.

G: What inspired you to look into it?

S: Well, one day I was walking with my buddy around the grounds over at Pico Rivera, where the B2 facility was. We were walking around at noontime talking about wouldn't it be neat if we could somehow build something that would screen an object so that you'd be looking through it. That's how I got started on it, discussing it with my buddy at lunchtime. And then, about a year later, I started actively working on it.

G: What was your immediate impression as to what the applications would be in daily life?

S: Well, my first thought was it would be used primarily by the military or law-enforcement personnel, but then, of course, there are a lot of commercial applications as well that could be used: security, hiding industrial facilities from view, consummative strategic assets. There are a lot of potential uses for it that could be used by the commercial sector as well as the government.

G: How could the military use this?

S: Well, mainly they would build a shield around a vehicle, an aircraft or whatever. A simple case would be, let's say, a tank, where you have a tank moving along like this, and you have one side shielded, so the tank is moving along here, and people are observing it from this side and all they see is the wall it's moving against or the background it's moving against.

G: The shield is blending with the background, and so it's not exactly invisibility, but more like supercamouflage.

S: Right. Well, it's a matter of degree and the scenario in which it's used. For moving targets, it's difficult. It takes highly sophisticated electronics, and you have to match the terrain, and it becomes especially difficult to match vertical lines. Horizontals aren't as bad as verticals, but matching vertical lines in the background is difficult. It can be done but it's very expensive. With stationary targets it's much easier. And distance is a major factor. The further you are away from the object, the easier it is to camouflage it. If you're a mile or so away, it's very easy compared to, let's say, three or four hundred feet. Then it becomes more difficult, extremely more difficult. And there's some physical laws that enter into the equation, where you can calculate exactly what degree of resolution you can get at certain distances.

G: What physical laws?

S: These are optics laws, basically, resolution of optics, and we did some calculations and came out with a magic number of seven hundred feet. If you're further than seven hundred feet you can do a very decent job of camouflage, even on a moving target, but as it moves in closer, it becomes extremely difficult. If you remember the movie *Predator*, the monster came out of the jungle, and when he came close, the jungle started moving around him. So that's what happens, basically. You get distortion, extreme distortion.

G: When I was reading your website, that's the first thing I thought of, the *Predator* movie.

S: Right. The movies do a very good job in portraying that. They have some good consultants no doubt working for them in optics. *Predator* is a very good example of one of the older movies where that law can be observed. As the creature got up in the top of the tree, he could be seen only if you looked extremely close. You could see a shimmering of light, a shimmering of the tree. Reflections on the leaves did not look normal, quite normal. But you had to look really close to see it. So that's the kind of distortion you get with this system.

G: So, if you're going to shield, say, an installation, a building, and somebody is flying overhead, how can you prevent it from being seen?

S: That would be pretty easy, the reason being that with a stationary facility you can depict the ground underneath, or whatever, or the surrounding forest—you can superimpose that on the cloth of the facility, and you can do a very good job of terrain matching so it's almost virtually impossible to see.

G: What about making it invisible to radar or heat emissions?

S: Okay. If the enemy is employing sophisticated sensors, then he may see that there's really an object there that does not match the terrain. But there are ways to get around that. There are some ways that you can fool these sensors, or you can lead them to believe that it's not what they think it is. A lot of this begins to get into classified areas when you start talking about countermeasures, counter-countermeasures and all this. It does start to get into classified material.

G: So if I were to ask you what certain countermeasures were, would you be able to tell me?

S: I can tell you in general, but when you get into specifics it may get into classified areas. If I get into numbers, if I start talking numbers with you.

G: Could you tell me in general?

S: Oh yeah, certainly. What were you thinking of?

G: In terms of countermeasures against, for example, if you have an installation and somebody is flying overhead and they notice that the terrain is not matching what they're detecting, how would you be able to fool them into thinking it's something else?

S: You would put a device on the surface of the object that you're shielding, and you regenerate a signal or some kind of a deception—deceptive signal—to portray that it's something different than what it really is. In general, that's what you would do.

G: Are you bound by certain classified...

S: Only on programs that I've worked on.

G: But...

S: Like B2, for example. But in general, I'm telling you things that are in the literature, the methodologies in general that I would be speaking about. For example, to shield heat, heat-generating sources, there are materials that you can put around the object, absorbent materials, that will absorb the radiation, or keep it down at a very low level, or even reflect it down toward the ground so it doesn't go toward the observer. There's all kinds of things you can do to obscure the radiation that's coming from the object itself.

G: Is this technology more speculative than operative?

S: No, it's pretty operative.

G: How long has it been operative?

S: These kinds of shielding techniques have been around a long time. I would say twenty, thirty years. There's methods, for example, of shielding heat coming out of the back of a jet engine, or diverting the heat so that it's deflected downward, or upward, as the case may be. If the observer is down below, they'll deflect it upwards where it's not very easily seen, or dissipate easier into the atmosphere. There's also ways of actually concealing jet contrails so they don't show up.

G: But in terms of the optical technology, of what you're doing in Project Chameleo, is that operative?

S: There are programs that I know are in operation. One is a painting scheme where an A10 has been painted with a type of paint that changes as you apply electrical signals to it. The color changes. It's called electrochromatic. And there is a project being tested on an A10 where you're looking up at the aircraft and the proper voltages are put on the skin so that it turns sky-blue, and, conversely, if you're above the aircraft looking down, they can change the color to look like the desert terrain down below. This is operative. And there have been a few articles on it, but surprisingly, they dried up. And not surprisingly, they dried up a lot. The articles are not seen anymore, so it's quite likely it's become classified now.

G: That's not as sophisticated as what you're proposing?

S: Not quite, but it's along a similar concept. The paint, the electrochromatic paint, has a different patent on it. It's based on the idea of blending into the background, but it does not take the background and project it on the skin. As far as my project is concerned, to my knowledge there is no direct evidence that it is being used, but the Army has released some information about their Future Soldier program, where they mention that it will have a skin that will blend into the background in a chameleonic fashion, but they don't describe how they're going to do this. But I can only assume that it's related to some kind of a sensor that would sense the background and then display it on the skin, on the uniform. I know they're working on this, but I don't know to what extent they've developed it.

G: You mentioned *Predator* before because you were talking about the building being stationary, but you just mentioned a scenario in which you can actually do this on a person.

S: Right.

G: Is that one of your intentions?

S: Oh yes, yes. I mentioned that, in my patent. I mentioned that it can be used to conceal men, vehicles or whatever.

G: And how would you go about... I understand the building being stationary, but when you have a person who is moving around, perhaps in a covert situation, how would you get him so that he's consistent in blending into the background?

S: You need to have a group of sensor cells on the side facing the background, and as the person turns, you have an array of these sensors, and you can have them blended in with display pixels as well. And there's even now, there is a type of cell, produced by NASA, that serves both as a sensor and a display. So you could use these kinds of cells. You could have hundreds of these on a uniform, and as a person turns, the background would be displayed in front, and through miniature computerized programs, you could program these so it would blend in pretty good with the way it should be looking from the view of the observer.

G: And what military applications could that have?

S: Obviously, for the Soldier of the Future, to blend into the background would be an asset. A terrific asset. And for, of course, spies or people who are operating in covert assignments, it would be ideal to not be seen.

G: You mentioned the Soldier of the Future, but what about the soldier of now? Do you think it's possible that... Dave Emory, who's a political researcher, once said that the private sector is always fifteen to twenty years behind the military in terms of technology. Do you think it's possible that something like this is already being done?

S: I don't think it's operational yet. In fact, the Army has told me that they don't have funding for this program yet. They told me that about a year ago, and the Future Soldier is a concept they have. They have people working on it, consultants, and they told me about a year ago that they do not have a program established yet, with funding and milestones and all, and the typical DOD development program has not yet been established. Now, they may be doing something on their own internal funds. That I don't know.

G: Is there anybody other than the Army that would be interested in developing it?

S: The Navy has shown some intense interest. I was interviewed by the Navy. The Navy came out to see me last March a year ago, and I made a presentation to them. This was a team of three people, Chief of Naval Operations Strategic Studies Group. And they were studying technologies that could be applied in the year 2025 in the fight against global terrorism. So the Navy is very interested and they told me that they intended to give Project Chameleo a good recommendation for the type of technology that could be applied by that time period.

G: This was March of when?

S: Last year, 2005.

G: And how did they find out about you and your project?

S: It was mainly through my appearance back in Charleston, at a Military Sensing Symposium, MSS. And that was in February, the month before, and a gentleman contacted me from the Navy about Project Chameleo and said they would like to come out and talk to me about it. So I said, "Are you going up to Charleston to the symposium?" And he said no, they could not make it there, but they would be coming out here. So they came out the following month, to Anaheim.

G: Tell me about the presentation. This was in Charleston?

S: It was a Military Sensing Symposium, and it was composed of Army, Navy, Air Force, and contractor personnel. It was a pretty multiservice-type group. And the papers covered a lot of areas, but one of the subgroups was camouflage, visual camouflage. So I made a presentation. I sent in an abstract and they accepted it. They were very interested in it.

G: How long was the presentation?

S: It only lasted probably thirty minutes, forty-five minutes.

G: There were a lot of people there, a lot of representatives of the Navy?

S: Quite a few, right.

G: And what were their reactions to what you had to say?

S: I had one very unsettling reaction. About midway through my presentation, the group moderator came up and whispered in my ear that I was revealing... that some-one in the audience told me I was revealing top secret information. And I said, "I don't think so." And this was following a remark that I made about the Future Soldier program, that they intended to develop a suit that would change in a chameleonic fashion. It was that phrase that prompted someone in the audience to say to the moder-ator that I was revealing top-secret information. I said I don't think so, and she said, "Well, see me right after the presentation." So I went ahead with the presentation and then I was hustled off into a room of about three or four security personnel, and they had me in there over half an hour questioning me, where did I get this information and so forth. And I said, well, I got it out of a public-release document from the Future Soldier program at NATEC. And the lady said, "Do you have that document with you?" And I said, "I sure do, it's out in the car." She said, "Let's go get it." So she walked with me to the car to get the document. We came back in and continued our conversa-tion with the security people there, and then they called in the lady who claimed that this was top secret information and questioned her on it. She finally had to admit that evidently it had been declassified. At one time it had been top secret perhaps, but now it had been released into the public domain. *[Laughs]*

G: Did they ask you if you were working on a top secret proj-ect? Obviously, you can't release top secret information.

S: They asked me if I was cleared for Top Secret, and I said, well, yes, I am cleared for Top Secret, but not on a pro-gram like this. I said, it's only on the program that I was formerly accessed to, the B2 program, so it has nothing to do with this. *[Laughs]*

G: Do you remember what branch of the military they were from?

S: They were from the Army. They were all from the Army.

G: Is it possible that when these type of people come and talk to you, they might be interested in seeing what you know about something that maybe they already have?

S: I think so, possibly. Or to get my slant on it. I know the Army really went out to bat for me to get me in that conference. It was not easy because my regular employer would not sponsor me. And this being a classified meeting, I had to get my clearance sent in. And since my immediate organization would not sponsor me, I had to go over their head and get my clearance sent in by headquarters. But then I was up against another obstacle. I still could not get into the conference until I had established a need-to-know, and my organization would not give me that because I was going under my own auspices, you know, my own personal program. So I contacted the Army and they went to bat for me. The head of the Army symposia group, I guess, wanted me there bad enough that he gave the go-ahead on it to the security people in the Army who gave me the need-to-know. So I was admitted to the conference.

G: Could you briefly mention who your regular employer is?

S: I work for the Defense Contract Management Agency, over in El Segundo. We mainly keep watch over the contractors around the U.S., and worldwide, as a matter of fact. We have an international division as well. And our job mainly is to monitor contracts and see that they're properly administrated and executed and we meet our schedules on time and under cost and so forth, or on cost at least. So that's basically what we do.

G: How long have you been working with them?

S: I've been with DCMA now about seventeen years.

G: Do they know about Project Chameleo?

S: My supervisor knows about it, right.

G: Do they have any feelings about you pursuing something independent of them?

S: No, no, as long as I do it on my own time there's no problem. There's not really any conflict. It's just... they didn't really feel that they should sponsor me to this meeting. They thought it was a conflict of interest from their point of view to say that I was authorized to go when I was not going under their auspices. And there is some credibility to that. The only thing is, I think they should have been a little more considerate and proud of the fact that one of their own employees had come up with something like this that the military is interested in. I think it's to their disadvantage to not give me the endorsement. Because a lot of agencies will go ahead and endorse you. Even on a private program, they will endorse you as long as there is no direct conflict with what you're doing on the job, and there isn't, there's no relationship at all. So I think they were a little shortsighted in their not doing that. I think in a way it was an embarrassment for them, to have the Army step in and sponsor me, you know, another agency.

G: Because ultimately it benefits them too to have the prestige of an employee who's been recognized in this way.

S: Right. I would think so.

G: Since 9/11 there's been a rise in surveillance. Is it possible that this technology will be used to monitor private citizens in the United States? Is that a concern?

S: Yes, it could be a concern, and our intelligence people could possibly utilize this to spy on our own citizens. I don't think it's very likely they would do that because they have other ways of infringing on our privacy that are more direct, like monitoring our communications and all that. So I don't think they have a strong need to conceal themselves from us visually in order to do something. I think a more practical application would be with security people who could be sitting, for example, over here on a tower somewhere, observing things, and this observation tower could be made totally invisible, and they could monitor traffic down below. They could monitor all kinds of criminal things going on visually by just sitting there watching what's going on. It's kind of an extension of video surveillance, I guess, except that in this case there could be operators in that tower. They could be looking out and visually seeing what's going on down below them. And this could add, let's say, a third dimension or fourth dimension to simple video cameras, because it takes a lot of cameras in many locations to see everything that's going on, whereas up on a tower looking down, you can see all kinds of things, and if people don't know you're there, it makes it even better.

G: So this could be used to counter possible terrorist cells.

S: Oh yes, I think so, absolutely. Counterterrorism. And the Navy, I think, appreciated that when I made the presentation on that. They could see the advantage of the fact that you could be invisible. In my presentation I showed how one of their new ships could be concealed and just sort of blend into the ocean—in fact, pretty close to the shoreline. And just for a little emphasis, I showed that, sure, I'm concealing the ship, but here's these two guys sticking their head up through the port hole, and they're sitting out there sort of above the water [laughs] and I had a little footnote saying, "Tell those guys to put their head back inside next time." [Laughs]

G: I wondered about that. Going back to the *Predator* scenario... if I have on one of these suits and I'm holding a coffee cup in my hand, does it include the coffee cup?

S: You could even eliminate the coffee cup, if you wanted to. You can do so much with the program. With a computerized system, you could screen out things that you wanted to screen out, or you could make it appear to be something that it really is not. You could make a human being appear to be an animal, let's say, or a tree, or... you could paint a portrait on the surface of the object of whatever you wanted it to be. It does not necessarily have to blend in with the background. If you had difficulty blending it, for example, make it a tree. That looks natural.

G: We've gone beyond just camouflage, but also to the use this could have—

S: Deception.

G: —in psychological-warfare possibilities as well.

S: Right. It gets more into deception, visual deception. Right.

G: So let's take Fidel Castro, someone like that, and you wanted to imbalance him, couldn't you use this to sort of slowly drive him mad?

S: Yeah, it could be used for psychological warfare. That's right. You could create a situation where a person thinks he's not seeing what he really should be seeing. That he's had too many drinks or he's on something. And, right, you could do things that would greatly disturb the psyche of a group of people.

G: And would it affect everybody in the immediate area or could you target a person in particular?

S: You can target a segment of people. To target one person is very difficult because of the optics of the situation. It takes a very... you would have to know exactly where the person is that you want to deceive. You'd have to know their coordinates exactly in order to do that, and that may be very difficult to achieve. I think in the future that we'll be able to do a lot of those things. We'll be able to pinpoint where people are by geographical coordinates, and then we can, in the simulation system, screen those people, or select those people to be targeted. I think that's in the future, but right now about the best you can do is target an area, a group of people in a… let's say, a quadrant, like that quadrant over there, let's say, forty-five degrees or so, one side or the other. To do much better than that right now is not feasible.

G: So if you had people in an enclosed area, you could just target that area, you could really bombard them with hallucinations.

S: Yeah. You could. You could actually create things that are so goofy that they would begin to wonder what's going on. Right.

G: Had you considered that possibility?

S: I think it's mentioned in one of my papers, but I'll have to go back and look if the psychological warfare aspect of it… I know it's been brought out by some other writers, and one of them in particular is a gentleman who's a investigative reporter. His name's Bob Fletcher, and he's put out a series of things about psychological warfare. He has a video called *Exotic Weapons of Mass Control*.

G: I'll have to look into that. Could you give us details about… I believe you're doing a FOIA investigation involving this technology?

S: Oh yes. FOIA. Yes.

G: The Freedom of Information Act.

S: Yes. I did launch a FOIA investigation, and the only thing I was able to determine was that the Army is doing some investigation. And it kind of stopped at that level. I was unable to get down below there and find out exactly what's being done by the Army.

G: You said the Navy was interested too. Did you initiate a FOIA request through the Navy?

S: Well, I went to an attorney and we went to all the services. Actually, we went to DOD, and then DOD eventually pointed us down toward the Army. And we were really unable to determine which documents should be requested, because the way the FOIA works, you have to identify documents, specific documents. You cannot identify merely a concept or an idea. You have to identify a set of documents. So if you don't have the names of documents, or numbers of documents, you can't do anything. So you can't get past that wall of secrecy, you know.

G: Ostensibly, the Act allows for freedom of information, but you need to know the specific serial numbers...

S: Right.

G: ... and you can't know that, so...

S: That's right, because you don't have access to it. Yeah. If you have, for example, a letter with a date and you know who signed it and you requisition that letter, then they have to either give it to you or deny it to you. And they can deny it to you under security laws and say it's classified. So they can stop you no matter what. But if it's unclassified, then they have to give it to you, theoretically.

G: So you need somebody on the inside to say, "Look for this specific document."

S: Right. Right. If you have somebody on the inside who can identify a document or drawing that's unclassified, you can theoretically get it through FOIA. Some of the UFO files have been obtained that way. And many times you'll see black marks throughout almost everything, so you end up with a page that's half black. *[Laughs]*

G: I've had those experiences. I believe it's even more difficult to get a Freedom of Information Act request through since the Patriot Act.

S: It probably is. They throw every roadblock they can in your path, on technicalities, to discourage you and to get you to give up on this. Put the burden on you to make a positive identification of exactly what it is you want in terms of a name, a number and so forth. So, like you said, it makes it very difficult unless you're on the inside.

G: Can you explain what prompted you to initiate the FOIA request?

S: Mainly I was motivated by patent protection. My patent is due to run out. I just renewed it again for eight years last year, so I've got another eight years to go, but if I don't do anything in eight years, I've lost it. It will go into the public domain. That frequently happens with patents. Many of them are high-technology or before their time. Nikola Tesla, for example, was way before his time. And many of these things, the cost of implementing something is so great at this point in time that it's not realized 'til after the patent has run out. I have some hopes, but I don't have great hopes, that I'll be able to achieve this before the patent runs out. *[Laughs]*

G: If this technology is being used now, shouldn't you have some rights to that?

S: Only if you can prove it. You have to prove the technology is being used. You have to prove that it infringes on your patent. The burden of proof is on you, the patent holder, to prove beyond a doubt that this patent is being infringed upon. It's very difficult. A good example is the man who invented the windshield wiper, the intermittent windshield wiper. He nearly lost it, but he finally managed to collect on that from all of the major automobile manufacturers, General Motors, Ford, Chrysler. All of them had to, in the end, pay off on that to the tune of millions of dollars. But the patent holder himself only got, like, I think, a quarter of a million dollars, and the attorneys got the rest. You know, that's what happened. His attorneys got the lion's share of it. And this was a long legal battle. I think it took ten years to resolve. And that was a simple case where it's a simple mechanical object that is pretty easy to prosecute relative to something like mine, which is more complex and very difficult to show infringement.

G: You can't even see it.

S: *[Laughs]*

G: Do you suspect that somebody is making use of your ideas?

S: I do suspect that the military is already working on it. The reason I say that is, a few years ago I was contacted by a company in San Diego and I met with one of their representatives and we had a long discussion on the subject, and I later found that they had a big contract with the Army and were developing something similar to this, but there was a wall of secrecy there I could never get past. It was a dead end.

G: This was in San Diego?

S: Yes.

G: And what year was this?

S: This was about... I think it was over five years ago that I had a conversation with those people and interchanged some correspondence, but it never led to anything positive. I think they were picking my brain rather than the other way around. The name of the company is... Starts with an S. I'll think of it and give it to you later.

G: You mentioned Nikola Tesla. I was just wondering how Tesla has influenced your work.

S: Only in an inspirational way. I've always been an admirer of Tesla, and as far as I know he didn't do anything in invisibility. But he had some revolutionary ideas about transmission of power and...

G: The uses of wireless technology?

S: Right. Yeah, he had a theory for transmission of power through wireless means, and he even sent power around the world. He sent a milliwatt or a microwatt, or some specific amount of power, sent it all the way around the world and it was a very, very small loss. Almost a negligible loss of power. But even with this kind of demonstration he couldn't get any funding for it. And I think there were huge consortiums against him, the power-line industry and all this. It would have really sabotaged the ground-transmission industry if you could actually implement wireless transmission of power. But I'm sure it can be done. And now, with satellites, it would be even easier to transmit power.

G: I wanted you to comment briefly on... In fact, we mentioned this earlier at the Lodge, the Japanese inventor whose name I can't recall.

S: Oh, Professor Susumu Tachi?

G: Yes. He's been in the news and he's received quite a lot of media attention.

S: Yes, he has. A lot of publicity.

G: What is your opinion of him?

S: Well, I believe I mentioned to you before, I think his idea is kind of hokey, and it's cumbersome and, to me, very awkward to implement, and it would be difficult to use in a tactical situation, because it seems to depend upon an external projector to be taken around with the person being shielded, or at a remote place from the person being shielded, to project onto the person being shielded. So I don't see it as a very practical way of concealing someone. It seems to me much more costly than my method would be, and cumbersome and... I just don't see how it could be taken seriously by the military. It might be used for advertising stunts or things of that sort, because you could have a projector sitting over here on a building somewhere, projecting down. That's quite likely. So I see commercial applications for his invention, but I don't see it as being very good for military or security purposes. It think it needs to be self-contained. Mine is a self-contained system.

G: Yours came first?

S: Yes, mine came before. Mine precedes Professor Tachi's. His claims don't really infringe on mine. I've read his claims, his actual patent claims, and it's based on a totally different concept. So I don't have any heartburn with him.

G: My question actually is that his device, his invention, is receiving a lot of media attention whereas nobody knows about you. His thing could not be used in a *Predator*-type scenario, making somebody entirely camouflaged.

S: Right.

G: So isn't it interesting that there's been a lot of media attention on that invention, which is so much less significant?

S: Yeah. He made a kind of a spectacular entry into the media, and it's probably through the Japanese media. It's my suspicion that they pumped it up and then it got worldwide attention. In fact, I was called by a reporter in Tokyo who published one of the articles on Professor Tachi. He called me from Tokyo and we had a long discussion, and he released later another article detailing some of the information I presented to him. So I think Tachi just caught a whirlwind of attention there and kind of rode the crest of it. But that's okay. It's... I'm not really out for media attention, although I think it does help sometimes to get the military to look at something. But I think now they've had the exposure, the military has been pretty thoroughly exposed to my concept, and it's only a matter of time and money, I think, before they will eventually do it. I'm sure they'll do it eventually. It's just that I may never ever get anything out of it.

G: You mentioned on your website that there are some commercial applications to your invention as well. I was just curious as to what those are.

S: Yes, commercial applications. I think it could be used extensively for advertising. There's a lot of schemes that one could use to promote products. In fact, I was contacted by an advertising firm in Florida, and they wanted to consider the idea for a store application where they would have a person who is at least partially shielded, and he'd be called the Invisible Man in the Store, and as you come in the store, the idea was to see the Invisible Man on Aisle 5, he has something for you. So you go to Aisle 5 and the Invisible Man is sitting there and he's partially obscured, or you see right through him and the background. And they wanted to use it as an advertising gimmick to call attention to certain products. And this is one application, but there are other commercial applications as well. A lot of security applications where you could build it into devices and in rooms where you have hidden cameras that looked like clocks. You could have all kinds of things where you concealed cameras and things that could be seen from a distance. Or you could actually conceal people in an area. It's probably more applicable to industrial security than it would be to advertising. Although there are some good advertising things you could use it for.

G: I'm assuming your initial idea was for military applications. The advertising thing came later.

S: Yes. Right. The military application was the primary motivation.

G: Did Col. Philip J. Corso[21] have anything to say about this kind of technology?

S: Oh, Col. Corso, I think, mentioned that UFOs had... Some of the papers that he had in his possession indicated that UFOs were able to sort of disappear, as if they'd go off into another dimension or something. And I think he mentioned also that inside one of the recovered craft that it was... It had kind of a translucent environment. As you got in there, it was like you were looking through it. You could see the outside like you were looking through the

---

21.   Author of *The Day After Roswell* (Pocket Books, 1997).

skin. But he may have picked that up from somewhere else too, because some of his critics say that he got this information through other sources. So it's a little... It's very controversial as to whether Col. Corso's presentation is authentic. I tend to believe him, I tend to think that he does not have anything to gain, or did not have anything to gain at his late age when he finally came forth with this, over age eighty. It's hard for me to believe that he would have risked his entire reputation, his family and all, to deliberately falsify information. It's just hard for me to believe that. So I tend to believe that what he says is true.

G: We've heard stories about the transistor and things like that being back-engineered from—

S: Right.

G: —the Roswell Crash.

S: Right.

G: Do you think that's possible?

S: I think it's possible. We know that one of our notable inventors is credited with the transistor, and it could be that he did invent the transistor, but then the Roswell findings could have accelerated the industry tremendously, even in spite of that. So I think it could have been a combination of both. I don't believe that the inventor that's credited with the transistor—I'm trying to think of his name right now, it's on the tip of my tongue—I certainly think that he is an authentic person and that he truly believed he invented the transistor. And I believe he did, as far as our own personal development is concerned. But it was long after this, I don't know the time span exactly, but Col. Corso claims to have found chips that looked like integrated chips *today*, and these could very well have greatly accelerated the development of integrated chips, transistors. So, yeah, I think it could very well be a combination of both. That creativity that we developed could be inspired by... You know, ideas have a way of being picked up in the ether too. A lot of times people will conceive of an idea at the same time, almost simultaneously. So if there are inventions that are already out there floating around, people can pick up on this. It's in the cosmos. *[Smiles]* Somewhere in the cosmos.

G: You're familiar with the writer Charles Fort? He wrote about strange phenomena back in the '20s and '30s: *LO!*, *Book of the Damned*. He had a theory called Steam Engine Time. That was simply that when it was time for the steam engine to be developed, it would be, no matter who it was who actually did it.

S: I've heard a little bit about that, but I'm not really familiar with his writings.

G: It goes to the same point. And perhaps it's true of how Project Chameleo was developed.

S: Right.

G: Do you think that the film *Predator* was perhaps getting the idea not from some fanciful source but from some actual application maybe, or some research that was being done in some area? Ideas have a way of leaking out into the film industry.

S: Right. That could very well be. The investigative reporter I mentioned to you before, Bob Fletcher, he has a film that I procured and played, and in there it has a little video clip of a person named Mike Burns, who supposedly is working on a suit for the military, and this goes back around the time that I got my patent, maybe a little bit later, I think, around the time that the *Predator* movie came out. I don't know the exact time period on the *Predator* now. But around that time period, this fellow named Mike Burns supposedly was working on something. And he used a clip out of the *Predator* in his demonstration. It was a film clip directly out of the *Predator*, so the *Predator* came out first, and then he had this little film clip in there. And Bob Fletcher picked up on this and it was in one of his tapes on strange weapons and things like this that the military is working on.

G: I think *Predator* came out around '87, maybe?

S: Yes, I believe it was around '87.

G: Do you remember when you saw the movie? Your idea predates that.

S: It's pretty close to the same time. I think *Predator* came out after I started thinking about it. But it is close to the time period that I started on it.

G: Going back to Col. Corso. I just had a question I've always been curious about, and since you're a professional, maybe you can answer it. The whole idea of the technology being back-engineered, you know?

S: Right.

G: I'm just wondering, how would that be... I'm imagining a scenario in which, say, a television set goes back in time and ends up in ancient Egypt, and they're pulling out wires and things, and they're trying to figure out how it worked. They would never be able to figure out how it worked. So I'm wondering, do you think it would be possible to back-engineer something so advanced with our technology in '47? Could we have even begun attempting to do something like that?

S: Well, Col. Corso claims that most of the projects they were able to back-engineer them, but with one exception, and that was the navigation system. It was like a headset that fit over the head of a person. It looked like what might be a plastic headset today. They could never figure out how it worked. And they finally gave up on it, apparently. Or at least they either gave up on it or it went into some very highly classified area that we will never know about, because he says they were unable to figure out how that worked. And when you put this headset on, you get a... after about ten or fifteen minutes you get a headache, start developing a violent headache. But they think that it was used to control the vehicle somehow. It was like a pilot interface between the pilot and the vehicle.

G: Yeah, I've heard speculation that the ship was almost run by telepathy. That there were no controls.

S: Yeah, by thoughts. Well, there's supposed to be some recesses for the fingers to fit into. And Col. Corso mentions that, and it's exactly what you see in the TV series *Taken*. They show the recesses for the fingers. Col. Corso claims that they had these pieces of the cockpit where the fingerprints were imbedded. They tried to make sense out of it but they never could. They thought it was some kind of an interface with the beams that were in there. And it could very well have been some kind of sensor, advanced sensor.

G: I've heard stories about a pilot interface in conventional aircraft that they're working on. Are you aware of something like that?

S: Well, I've heard of doing it with the fingers, by finger controls, like that, but the concept that *Taken* has and Corso mentioned was that apparently it was through the mind of the operators, together with this headband thing, that the headband played into it somehow and they were never able to figure that out.

G: One more question about that, and that is, you've been cleared for Top Secret projects... I meet people who will say, "You know, the government is really inept, they can't keep secrets, it's not possible to have a conspiracy that large." But you've had experience in Top Secret. Would you say that's true?

S: Well...

G: In other words, is it possible for the government to keep secrets?

S: It's possible for them to keep a deep, dark secret, and they do. However, sometimes they *want* to leak it out. They want a controlled leak. And I firmly believe that the UFO phenomenon has been leaked out in a very controlled manner since about the 1950s and '60s. They started releasing to the media the ideas of the Roswell findings, and I really believe that much of this has been stimulated by leaks from the government. And that it was deliberate. The reason I say that is because there's just too many coincidences and too many connections, and a lot of executives from the CIA, for example, have moved back and forth into the movie industry. It's like a revolving door between the CIA and the movie industry, and the media, the national media. And the whole idea behind it is that it's a form of conditioning and control, that if you condition the people that this is the way it's gonna be, then when it finally comes, they're not going to be shocked out of their mind, you know. And I think it's worked pretty well, because right now, I think, if we saw UFOs and they really did land, we wouldn't go into panic like we would have back in the '30s, when Orson Welles came out with his thing. I think today that people would be a lot more mature about it, because I've heard the polls say that 70 percent or more of people believe in UFOs, and so if they did appear, and with all the movies we've had and all the conditioning, I think that it wouldn't be a great shock. There would be some people maybe that would commit suicide. I think there'd be some people who would go bananas. But they would be some of the less stable people in society. In answer to your question, yes, I think the military can keep a deep, dark secret, and I think they have done that to a degree with Roswell. But they've also leaked it out, which is kind of a paradox.

G: Do you think these leaks might be misinformation or disinformation?

S: The advocates of UFOs will say that it's leaked out deliberately to condition the people.

G: Yeah.

S: The people who are against UFOs say that it's disinformation. So it's all a point of view. But I think I'm on the side where I think they're really trying to condition people so they won't be shocked.

G: Time-release aspirin.

S: Yeah. Yeah, right.

G: And Col. Corso, a quote that I remember to this day, he said, "The disclosure is the cover-up, and the cover-up is the disclosure." In other words, you release the truth in the form of fiction.

S: Yeah.

G: And in the form of fiction people get used to the ideas, and also, at the same time, if somebody does know something, says something publicly, people will think, "Oh, he got that from *The Outer Limits*."

S: Yeah, right, right.

G: And so it works both in conditioning people, getting them used to the idea in the form of fiction, but also in covering it up.

S: Right, right, exactly.

DF: Do you really feel that they think we're so stupid that they need to condition us?

S: I don't think they think we're that dumb. I think there are segments of society that they're extremely concerned about, and one of them is the stock market. They don't want this to become public knowledge because you would really get violent excursions on the stock market that could affect many millions of people, including a lot of millionaires and billionaires. They don't want that to happen. And that's why they, I think, are playing this game of this controlled release to... The skeptics, who are most of them in the financial world, they're very hard-headed, they see only the bottom line. So if they don't see an effect on their stock, they're fine, if they think the stocks are a good investment. But if it were to be revealed that here's an invention that could shake the world... Just to be laughable about it, the capsule that you drop in your gas tank, you fill it up with water and put in a gas pill. If there were such a thing as that possible, this could shake the world. They would not want anything like that released. Anything that is that dramatic. And that's why, I think, today a lot of the energy, the alternate energy, is being played down a lot, discouraged, not financed.

G: Speaking of disclosure, there's a reason I brought Dion here.

S: Oh?

G: He has a very interesting story which will intrigue you.

DF: Okay. In July of... what year? How many years ago? Three years ago?

G: It was July of 2003.

DF: In July of 2003.

G: In fact, the same week I went through the third degree.

DF: I've been working, I've been living in San Diego almost at that time about three or four years. I had a fiancee, a good job, etc. That relationship kinda went south.

S: Mm-hm.

DF: And, you know, I was working in a pub bar and restaurant, and then I injured myself, broke my ankle and stuff. Because of being out of work and stuff, I needed a roommate. And not using the best decision-making, depressed from the breakup and all this stuff, I'd chosen this guy that I later found out was on drugs. From his experience in the drug world, he ended up befriending this kid who was a deserter from the Marine Corps. Desertions were worse than AWOL, because you could be shot on sight.

S: Oh, really?

DF: This kid was twenty years old and in the high-tech end of the Marines. And when he deserted, he'd stolen over a half million dollars' worth of night vision goggles, a laptop and a 9 mm taken off the body of a dead Iraqi general, supposedly, and other stuff. Brought that right into my house.

S: Wow.

DF: Yeah. This kid was there three days. After this horrific experience, suddenly I'm the target of investigation from the Navy Criminal Investigative Service, Department of Defense. I have intimate experience with your technology. I know it works for a fact. I know people were in my house who pushed me aside. I could feel they were there, because I could sense something, but I couldn't see them.

S: Really?

DF: Swear to God.

S: Wow.

DF: I felt them moving...

S: Really?

DF: Yeah. Pushing me away. And also I was wondering, with, like, we were talking about this earlier, right when I met you, about the cars. I was on the beach and I swear there was like vehicles moving around the sand where you see the tracks, moving, going, but I could not see the vehicles.

S: In San Diego?

DF: In San Diego. It really just cracked me up when you mentioned the San Diego company, because it did seem that this was definitely connected to the military, and when I asked them about what was going on and who was watching me and da-da-da, they would say, "Well, we're not watching you." And I felt that the private sector was totally involved in what I've gone through, to basically psychologically torture me. And they were using stuff, like electronic warfare-type stuff. I mean, my kidneys were bleeding.

S: Really? Wow.

DF: So, yeah, they do use this stuff on Americans. You've talked to me all day. I mean, you know, I'm a balanced person. I think, anyways. And him, he's a professor, he's my best friend, he's known me for almost twenty years. This is just like an experience that... It's three years ago and I'm still not over it.

S: Wow.

DF: And psychologically, you know... But I'm like so, I mean, like I told you earlier, I'm the one who found your website, because I knew that this was a reality because of the experience. And then finally to meet you... which I'm glad I have. You're an amazing man. I'm really pissed with not only what they did to me, but they've just stolen your technology. And that sucks. That's not right. And it's like I do feel, like you said, it's going to be hard to prove, especially with something that's invisible. How do you prove that they're using it? But I do know of people down there that have been victims of it and stuff, and it was particularly the Navy Criminal Investigative Service. The Navy seemed to be really involved in this.

S: Really.

DF: I think, yeah...

S: Hm.

DF: And I think what you said was that they're the ones that had shown such an intense interest in your work.

S: Right.

G: You might want to back up a bit and mention that they actually arrested you.

DF: Yeah, they arrested me, and I was in jail for six days...

G: Because of your connection to...

DF: To this guy. But I didn't do anything wrong.

S: This was the guy who brought the stolen things into your house?

DF: Yeah. Where did it go? And they just refused to believe that I didn't have it. It was more of a denial thing, and they wanted to know, and I'm like, look, I didn't have anything to do with this. I didn't take this stuff. Your people, the people you have in the military, are the ones that are responsible for this, not me. So I'm a free man today. I was in jail for six days, I never saw a judge, never saw an attorney. And then, for about eight months or so, after I got out of jail, it was from July to...

G: To January.

DF: Yeah.

G: Constant surveillance. I have to say at first I thought he was going crazy.

DF: Yeah. Talking about invisible people and whatnot...

G: I should point out that we had never heard of anything like this. However, he started describing to me the symptoms of what was happening, and I'd get off and talk to somebody else, and we'd start mapping out, back-engineering the symptoms. How could this be? And we thought, could it be a full body suit? And everything you just described is what we were trying to figure out.

DF: Actually they're using a combination of your technology and something else, because I would be in my room with a mirror and like moving the mirror back and forth. You know, a mirror that you use to look underneath your car for brake jobs and stuff? It's a little mirror like this? I would sit there and I'd move it all around, and I would catch them, and I would catch like a part of their...

S: *[Leaning forward with intense interest]* Really?

DF: Yeah. And then they would move, jump out of the way and stuff. And then there was, like... I think they put things in... Did you ask him about Fakespace already?

G: No, I didn't ask, but he did mention something about... when you were talking about how you can make something look like something else, you described that to me.

S: Right.

DF: Yeah, that was the most... Yeah, because there would be people looking like they were morphing into trees and stuff.

S: *[Nodding]* You could have been used as a test subject or something.

DF: Yeah, that's what it was.

G: That's what we eventually realized. And eventually... I was talking to a friend of mine, I was describing this to her, and she's, like, this guy's crazy. And I said to her, I said, Why not? He's within the same area. They think he's some kind of terrorist stealing stuff. Why not? They think it's justified.

DF: They thought I was a piece of crap. And it's just like...

S: Yeah. And they could pick on you because...

DF: Nobody would believe me.

S: Yeah. You know, I just thought of the name of the company. SAIC.[22]

DF: SAIC.

S: You'll find them, I think, in the directory, unless they've changed their name or merged...

DF: No, they're there.

S: I think they may have stolen my idea. I really do.

DF: I know for a fact...

S: The reason I think that is because I had a meeting with this guy from SAIC. It's been about seven years ago.

---

22. The next day, on 3-12-06, Richard sent me a follow-up email in which he confirmed the corporation as "Science Applications International Corporation (SAIC)." He then went on to report, "In my archives I also have some correspondence that I conducted with the company over ten years ago."

DF: Yeah.

S: Or more. Let me think back. It was more than that. Actually it's been twelve years ago.

DF: My goodness.

S: Yeah. Twelve years ago I met with this guy. And around... Maybe only ten. Ten years ago I met with this guy. It could have been twelve. But anyway, we met in a restaurant here in L.A. He works for SAIC, or he did at that time. And I've got all the records on this meeting.

DF: Great.

S: And he was amazed that Dr. Schweizer and I, my associate, Dr. Felix Schweizer, and I had come up with the same things that they had been working on, in isolation. He said, "It is amazing you've come up with much of the same things we have." And then I explained to him that I'd been in contact with Dr. Thomas Hafer, back at DARPA, Defense Advanced Research Projects Agency, and he said, "Yeah," he said, "that's the right person you should be dealing with. Dr. Hafer." And so we had a lot in common. He was telling me a lot about what they were doing, that they had a contract with the Army, they were doing things with the Army. But then, later, I wrote a letter to... almost immediately I wrote a letter to the president of SAIC and I suggested that we enter into some kind of a mutual collaboration group, and that I could license them, because I have the patent rights. And I got a letter back saying something like they would consider this and get back with me, and that was the end of it. Then I sent a follow-up letter and I never got a response. So I came to the conclusion that these people were already doing it.

DF: Yeah.

G: These people sound like vacuum cleaners. They come in, they suck up information—

DF: Yeah, and they just go.

S: Yeah. Now, we were doing it so much near the same time frame that, legitimately, they might claim they came up with it independently. And it's possible. But I have... I have the patent on it, but now it was brought out to me by the patent office that there are patents that are classified. They're not in the public domain. They've not yet been released. They're reserved for the military, and it could very well be that SAIC has that kind of a patent pending. It's called "patent pending," a release from the military because you cannot have a secret patent. In other words, you can have a secret patent pending, but when you file a patent... it's put in a public book, the patent office issues a book, and so it cannot be classified at that point in time. So if the military is using things, they can do it in a clandestine atmosphere that's protected by U.S. patent laws. In other words, the military is allowed to use anything they want. But they can use my patent if they want to. They have the right to use my patent. However, if a company starts producing cloaking systems, that's where I can step in and sue the company.

DF: Yeah. See, I think...

S: I cannot sue the government.

DF: You can't sue the government.

S: That's because they have the right to use any patent in the world. They have unlimited use rights on patents. But that does not mean they can manufacture and sell them. They can't do that. They can't manufacture and sell them. So what they do is, usually they go out and get a company to develop and manufacture something. So that's when you have to step in and sue them for patent infringement.

DF: Yeah, that's in this paper I have here, which is the Future Warrior Project, or the Objective Warrior Project.

S: Oh yeah, yeah.

DF: Did you ever talk to a Captain Jean-Louis D'gay?

S: No.

DF: He's quoted in this article I have here. It says, "But the most fantastic ability of the smart uniform will be invisibility. The Future Warrior will blend into his surroundings like a chameleon."

S: Yeah, yeah, right. Exactly.

DF: Says D'gay, "The U.S. Army has always lived by the motto 'We own the night.' Now as opposed to just being a shadow, we make the soldier invisible. One of the biggest revolutionary events that we're looking at is this chameleonic camouflage," which I find it ironic they use your terminology.

S: Chameleonic, yeah.

DF: "Nanotechnology gives us the ability either to change or use camouflage pattern in a textile in a moment's notice or even create a mirrored effect." And it talks a little bit about *Predator*, and it says specifically, "This is no fantasy to the American military. Experiments with the technology have apparently been so successful that the U.S. Army immediately yanked the development away from MIT and completely classified invisibility research."

S: Oh, they have. Okay.

DF: Yeah. It says, "In a world where special operations forces and stealth are increasingly important, invisibility may become the new atom bomb. Countries that have it will be powerful while those without it may be forced to submit."

S: True invisibility would give someone great power.

DF: It's such a great... It's almost an "absolute power corrupts absolutely" kind of thing.

S: Yeah.

DF: It's a very scary thing. When the head investigator... The way I'd handle the situation, is I'd go get my hose and go out and...

S: Did you do that?

DF: I would get food, I would get food out of my refrigerator and just throw it or...

S: You could see the object?

DF: I could hear. I've heard somebody, an invisible person, fall out of a tree outside my house.

S: One good thing, if you have a spray can on hand...

DF: That's what I was wondering. Maybe something with ink in it, like you spray something with ink and then you see your ink moving away.

S: Well, in *Memoirs of an Invisible Man*, that's another good one, where he's running through the rain and you can see the water running down over him.

DF: Yeah, in the movie that's really good.

S: So that's where invisibility is really vulnerable.

DF: Yeah. And also vulnerable to an AK-47.

S: Right, right, right. If you know where to shoot.

G: Yeah, that's why it would only be used in a covert situation, probably one person going in.

S: Yeah. Because once you detect it, then you could be shot.

G: When I was talking to my friend and I was telling her about this situation, she said, "Well, this doesn't make any sense. Why wouldn't they just test something like that on the Marine base?" And I said, "Because that scenario is not random enough. You're dealing with somebody who's random and you don't know how they're going to react, you can more accurately test what the vulnerabilities are." You know what I mean?

S: I truly believe they do sometimes pick on people.

DF: Yeah, they do. With the Tuskegee experiments...

S: People who are vulnerable, I think.

DF: There were those black guys down South who were prevented from being treated for syphilis and stuff. And they've tested chemical weapons on their own people. Just reading in the paper yesterday, we were looking at an article about Chechnya, and people coming down with a mysterious illness, which sounds like chemical weapons are being used on them. Respiratory problems, people going into uncontrollable spasms. It just sounds like that's what they're doing. They do. That's what the military has always done. But it's an honor to talk to somebody who actually has created this revolutionary technology that is changing the world. It has already. That's a fact. It's almost comical. I threw everything I owned into the van and I just had to get out of there.

S: This is really fascinating.

DF: And I just drove, and I was followed all the way across the country. And every time I'd get so far and they would leave me alone. But as soon I hit a military base, another group of people would start following me.

S: No kidding.

DF: There's a lot of military bases across the country.

G: You should mention the time when they actually admitted this to you. The time with Lita.

DF: Yeah.

G: The lead investigator was Lita A. Johnston.

DF: From the Navy Criminal Investigative Service, Miramar.

S: She revealed something to you?

G: Well, it was a comment.

DF: Yeah. She said I made her guys laugh a couple of times.

S: Really?

DF: So I would go out there and... you know the cap and gown you wear for graduation? They were trying to gaslight me, drive me crazy, so I would feed into that and wear the cap and gown and go out and throw cheese everywhere. And then go back and take the broom and sweep it up real fast and go back inside and stuff. Because I knew...

S: She actually identified herself?

DF: They were asking me, when they'd come to the house to arrest the kid, when they found out where he was at, I was totally cooperative with them. I'm, like, look, the stuff that you think is in my house, you're more than welcome to… Look, I have a roommate. I'm not responsible for the stuff that you might find that could get me in trouble. Drugs. I didn't want to get in trouble for somebody else's crap in my house.

S: Yeah.

DF: So she signed a piece of paper saying they were only looking for the night vision goggles, and we agreed upon that. And then they arrested me for the other people's stuff that they found in the house.

S: What kind of things? Night vision goggles and what else?

DF: Laptop and a gun and stuff, but I really think that… These were night vision goggles off an F18 fighter plane. You see roughly three times as far in dead-dark. They're incredible night vision goggles. And I think maybe this stuff might either work in conjunction with the invisibility or just…

G: This is something we were speculating about, because we couldn't figure out why they're obsessed with these goggles which you'd think they could get anywhere. We were trying to figure out if there was something special about them that tied them into your technology.

DF: Maybe that you could see the invisible people with them on or whatnot. I still don't happen to have the answer to that question. You guys were talking about the FOIA documents. Just yesterday I filled out my request for the FOIA stuff.

S: Oh, you did?

DF: Yeah, because… And this totally relates directly to you and your stuff, because if they were using this on me, it might be in FOIA. I don't know.

S: So did you have specific documents to...

DF: No. I'm waiting to get it back and stuff. But the great thing about it is the name of the company, because if we can tie this back to a private company, you get a paycheck, because this is part of your technology. And there's no way around that, I don't think. Because I experienced it, and I know I experienced more what you talk about on your website. I experienced that. When I read about the Japanese guy and stuff, I knew, nah, that's not it, but then, when I found yours, I'm like... I emailed Robby, going, "This is it. This is what they're using."

S: It could very well be SAIC doing that.

DF: Yeah. It sounds, if you talked to these people, that that's what it is.

G: You didn't finish the story. Once he had the food fight with the Feds, Lita Johnston asked to meet him at a café, and he went there and had lunch with her and her superior.

DF: Yeah, I told them, I said, "Look, I'll steal this stuff back for you." There was a woman in my house that day that had three children, and they weren't in the house that day, but I knew how this would impact her life if she got in trouble, so I refused to tell on anybody that was in my house. They got the kid that they wanted, they got the stuff out of my house, I figured that's well enough. So I'm not a snitch, I'm not a rat, and I refuse to tell on anybody. So that infuriated them. That just made me the enemy in their eyes. But I told them, "Look, I'll get the stuff back for you if you pay my rent, because now I'm way behind in the rent because you won't leave me alone."

S: *[Laughs]*

DF: You know, and stuff. She laughed about it. She wasn't, like, a totally cold-hearted person. But her superior just looks at me and goes, "Enjoy this food, it's on me." And I go, "This is my tax money. You're..."

S: Enjoy what food? What food is that?

DF: They bought me a bagel.

S: Oh, I see.

DF: And an orange juice.

S: Oh, I see. *[Laughs]*

DF: And then he just said, "You have the stuff. You have them." And I'm like, "No." And I don't. I'm not lying now, I wasn't lying then. But they refused to believe it, just out of... I don't know, out of ignorance, just denial, I don't know. But at that point I knew I was just in a sinking ship and I had to get the hell out of there.

S: How long ago was that?

DF: It was three years ago.

S: Three years ago.

G: During that meeting he made a comment. He was like, "Listen, I don't want to have any more food fights with the Feds," and Lita Johnston smiles and she goes, "That made us laugh."

S: The one time she admitted it.

DF: There was another time and when I ran out with a bag of flour, and I realized as I'm throwing the flour... This was not too long after the anthrax thing, and I'm like, "Oh, this isn't a good idea." So I'm, "No, no, no, I'm sorry. This is just flour!"

S: A bag of flour...

DF: That wasn't smart. I realized just as I was in the motion of doing it and I stopped. But there are friends of mine, two people that I am concerned about that live down there. I hope they're alive, because my dad, who I told you is a retired police officer, and Rob, he's a school teacher, were both concerned for my life. They thought that any moment I could be shot or whatnot. Because these were kids with guns in my bushes. This technology has gotten into the hands of really irresponsible people. And I knew they were in the same room with me and stuff, I'm like, why don't you rob a bank or something, go to a girls' locker room.

S: Yeah.

DF: Why mess with me? And so she commented that I crack them up and stuff. But this is going to be impossible to prove.

S: How long did they hassle you like this?

DF: For about eight months.

S: Eight months?

DF: Yeah.

G: The surveillance also was less technological. They would just follow him.

DF: Yeah. I'd have a parade of people behind me.

S: Wow.

G: Mention when you threw the beer.

DF: Oh yeah. I was outside a business, and this guy who... I ran across the street real fast and I think he thought I was running away from him, so he comes running after me. I dead stopped and he's coming. And so he still has to continue to run, because it would have looked weird if he would have just started walking. So I just threw beer on him. And any person who was a stranger would have gone, "What the hell did you just do that for?" But since they were given an order not to confront me in any way, there's a lot of different situations where I would do that...

S: You could get away with something like that.

DF: Oh, ridiculous behavior. I was in a grocery store where I knew the guy was following me, and he was behind me, and just suddenly I got out of line and got behind him, and just like with a tin foil thing, stuck it on his ass. And anybody would have been like, "What the hell!" He just stood there, going "r-r-r-r."

S: He just ignored you?

G: They must have said...

DF: Don't engage.

S: Don't have any engagement with him.

DF: She mentioned... I forgot the term she used, but she mentioned that. But then there is a mechanic about two blocks away from me, who later, after this initially happened, I met. A general in the Navy came into his shop and says to him, "We want these back. If you know anybody that had anything to do with it, we're going to get these things back."

S: You mean those things that...

DF: Yeah, whatever was taken. He said, "I'm going to get my stuff back." Like this was his property. This is the mentality that they had, like they were personally ripped off. And this is... The Navy is in charge of investigating my case. And there's always that rivalry between the Navy and the Marine Corps, and this is just feeding this fire between these two branches of the military that now have the ability to become invisible. It's almost childish...

G: Mention the comment in the bathroom.

DF: Oh, this. I'm across the country, forced out of Minnesota by cops that were watching me in Springfield, Minnesota, where a friend of eighteen years asked me to leave because they came and talked to her husband. And she's like, "I have a kid, you can't..." And I'm like... It was just horrible. I left. They drove me literally to the border. And I talked to Robby on the phone from a truck stop for, like, three hours. It's five in the morning now. I'm in the bathroom, and the guy comes up to me and goes, "Just give it all back and this will go away."

S: Ugh.

DF: And I turned to him, and just as he says that, another person comes in the bathroom and then he leaves, it's all, like, what the hell? It was just like totally out of a movie, like...

S: Wow.

DF: ...being trapped in James Bond hell.

S: Oh wow.

DF: I'm getting better now. I'm still having trouble sleeping and whatnot.

S: But these invisible people came into your room...

DF: In the house.

S: How many times did they do that?

DF: I would say half a dozen times that they invaded my space in the same house, that I know of. There's other... do you know the concept of Fakespace and how your technology is used to make rooms seem smaller or larger?

S: Yeah.

DF: Yeah, I've seen them do that. And they would do ridiculous things too. They used this technology, I think. It's the same technology and it just, like, widened my house by a few inches.

G: People noticed it, right? They came in and said, "What the hell?" The room's bigger.

DF: The room's bigger.

S: Yeah, you *can* make it bigger. Absolutely.

DF: The only purpose they did that was to just make me bonkers, right? I mean, that couldn't have had any practical purpose other than that.

G: Didn't you tell me, "I look out the window and the scenery seems different now"?

DF: Oh yeah. And, okay, also this. Can they make it look opaque, and there would be people in a car and they make it look like the windows... like there's nobody in the car?

S: Oh yeah, sure.

DF: Yeah. I told you, eh? I knew they were sitting right out in front. They followed me in and then the car was parked there, and I had just gotten a banana and a chocolate milk, and I could see them, really see them. I just sat there and just ate the banana. And they just sat... Four people going, "R-r-rr." And I took my sweet time and just went about my business. But it was really, a really bizarre experience. I'm really worried about... You know, I was worried... Nobody's messed with me in a long time. It's been over a year.

G: Since you went to Kansas.

DF: Yeah. It's funny. I moved to a small town in Kansas and they left me alone. They didn't seem to like Kansas.

G: How many people in the town?

DF: Oh, a hundred and twenty maybe.

G: I was thinking maybe the town was so small, they couldn't do any of this.

DF: Oh yeah, it was. They couldn't pull anything off. It was so small...

S: They'd be observed too easily maybe.

DF: Yeah. And I was... my next-door neighbor had a hunting club, so there's guns going off continuously.

S: When these invisible people came into your room, did they sometimes show up like you could see part of them, or they glimmered?

DF: Yes, yes.

S: The surface was kind of—

DF: Yes. Yes. There would be this weird—

S: It was kind of a vibrating or sparkling?

DF: Yes. I've suffered from migraine headaches since I've been nine years old. The migraine headaches cause... Jesus, I just forgot the term they use. But it caused...

MO: Auras?

DF: What's that? Auras? Yeah. And an aura has the same kind of sparkly type thing. So at first I thought it might be that. Like a sparkly mirrory type effect.

S: Yeah.

DF: And it's like a suit of mirrors. I've seen, like, a suit of little old mirror-type things.

S: Yeah, that's what happens when it's not perfect and you have trouble getting perfect matching.

DF: I'd like to see this sometime. If you could show us a demonstration, if that's ever possible, I would love to see what you're able to do.

G: And then to see if it looked like what you saw.

DF: Just to see it, period. And to confirm that. Like I said before, I'm… I had aspirations and stuff that were becoming realized. At the same time all this happens, so all of it is on a back burner. But I still want to become an investigator and stuff, and hopefully I could use all this as the launching pad to look more deep into it, and go down to San Diego and do a little bit of footwork, with these people with SAIC?

S: SAIC?

DF: Yeah. And stuff and…

S: I'm not sure what SAIC stands for. I think S is Systems and I is International, but the A part floors me a little bit now. I can't think of it now.

DF: There's a lot of companies in San Diego doing covert-type research.

S: Systems Research might be operating down there.

DF: As I was on the Internet investigating all of this out of San Diego, I found American Technology Corporation. They do more of the sound stuff. Because they were also assaulting me with sound stuff. So they're able to now have speakers that are completely directional, where I could direct a speaker at them and you could not hear it. So it sounds like there's people talking to you inside your head kind of. And there is the total commercial applications for people in bed next to their spouses who're watching TV. You watch TV without disturbing the person next to you. And there's also military applications for this.

S: Right.

DF: I was able to, or *we* were able to, take my symptoms, from what I described to him, and link it directly back to this company, American Technology Corporation.[23] And we wanted to be able to do that with the invisibility...

G: There was almost no information on this stuff in 2003. I would search the Internet all the time and there was nothing. I couldn't find...

S: 2003?

DF: And then this kid popped up. I didn't know he was a professor. Tachi?

S: Tachi.

DF: Tachi, yeah. Because he was pretty young, I thought. But it could be... I have pictures from the Internet that show him with a brick and he's going like this, and it is real clunky. In fact, it says that it's very rudimentary and it says that you could use digital video cameras to simulate the invisibility thing somehow.

S: Yeah.

DF: And that anybody could do it.

G: Mention the silhouettes.

DF: Yeah. There were times when I thought that I was seeing a shadow, from them moving. Is that possible?

S: Yeah, casting a shadow.

DF: I also saw things that, like, looking out my window, I thought I was... The one and only time I called the police on myself. Because I was looking out and it looked like people were jumping the fence to come into my house...

S: Yeah.

DF: And I panicked and called the cops and the cops were at my front steps in two seconds. They were there. And it's like I don't know what was used exactly, but I would look out my window, oh, continuously, I would look out my

---

23. Update: In March of 2010 the American Technology Corporation changed its name to the LRAD Corporation.

window and I could tell what I was seeing was not what was out there. It was just an image of what was out there.

S: Oh.

DF: I think this was maybe a technology that had more to do with something that's inside my head.

S: Now, these people that came into your room, they should have cast a shadow unless they had very sophisticated shadow eliminators.

DF: See, but in the house, the shadow thing was like he was dimly lit and stuff. But I was in San Diego and there was stuff outside the house, so I do think I've seen, like, shadows moving and stuff. And so I don't think they've one hundred percent—

S: You can eliminate shadows, but it takes energy to do it.

DF: Oh, static electricity disrupts your vision, and they were able to... There was times when I felt like my vision would be disrupted, then they would implant an image, like either outside of me or possibly in, like, you know... I'm not sure. It's crazy.

S: I wonder if they're doing anything with holographics.

DF: Yeah, that's what I meant.

S: Because you can do a lot with holographics.

DF: Yeah.

S: But you need to have an external projection device.

DF: That's with the things on the windows. It seemed like they were just displaying a hologram of the image of the outside. That's what it was, holograms. And I'd do, with my mirror thing... I have a memory of somebody with a box this big, and we had gotten stuff off the Internet that unfortunately isn't in your package, which talks about, like, using the invisibility technology for surgeons for doing surgery. They could see through their own hands to...

S: Yeah. That's right. I heard about that the other day.

DF: That is just, I mean, that's your paycheck. That's your money. If they're taking this technology which you patented and they're going to use it in surgical application, that's completely—

G: What you mean is you actually saw one of these guys for a moment and he had a box—

DF: Yeah, it was up above and stuff, you know. I would have to take a mirror and go like this, and then I could catch a quick glimpse of it like above my head.

G: Does that make sense, that you could see something in the mirror if it was using the technology you're describing? Is it possible that somehow—?

DF: To catch it at a different angle in the mirror, it just...

S: Yeah, it would distort it, yeah.

DF: Being able to see the invisible person with a mirror? I mean, I know it's, I know for a fact that—

S: Yeah, you could, because the mirror would be outside the range of the...

DF: Projection.

S: ...of what they're trying to shield. They're shielding you but not the mirror. So if the mirror is down here, yeah, that makes sense. Because they developed... Designing these systems, they tried to take the most obvious solution. They're not trying to solve everything at one time.

DF: I came at them from every different angle. Food, mirrors—

G: Which is exactly what they want.

S: They might have been... They might have been testing... Yeah. *[Pause]* Using you as an object to react with to see what... to further develop their methodology.

DF: Yeah.

G: You see, they would rather use the stuff on Dion first rather than start out in some dangerous, covert situation in Iraq. Or test it on someone prominent who would draw attention to their experimentation.

S: Oh yeah, they picked somebody they felt they could discredit.

DF: Yeah. Yeah. Which is definitely...

S: You know, they used another guy named Bob Lazar that way.

G: That's true.

S: Have you heard about him?

G: Sure, sure.

DF: Who's Bob Lazar?

S: Bob Lazar. He was used for a disinformation scheme and... I believe that he really did work over at Area 51. He revealed the things going on over there. He almost got killed.

G: They had him under surveillance too.

S: Oh yeah, for quite a while.

DF: Is he the guy who said all that stuff? He said he'd worked there but there's really no...

S: No, no, he clearly worked there, because he knew specific things about S-4. However...

G: There is no official documentation on him.

DF: Yeah, I've read about that.

S: Well, they could easily destroy all that.

G: That's the thing. People have gone back to the university where Lazar said he went to and they couldn't find any record of him ever having been there.

DF: He was there. Yeah, he talks about seeing bodies and stuff like that. That's the guy, right?

G: No, he never saw any aliens. He only dealt with the technology end of it.

S: He saw the vehicles.

G: Just the vehicles. He had a very specific function, because everything was compartmentalized, just like the Manhattan Project.

DF: A need-to-know basis...

S: Yeah. He had a very narrow area where he worked.

G: And then later on they nailed him for helping build the surveillance technology for the prostitutes—

S: Right.

G: He couldn't get a job anywhere, so he got a job making surveillance for this whorehouse in Nevada.

S: Yeah.

G: And then they busted him for it.

DF: Jeez.

S: He was ready to go to Japan to talk to the Nippon, NHK, that's the big outlet in Japan. He was going there for an interview and he said he was down at the airport ready to get on a plane and they shot at him and he had to abort the flight because...

DF: Every time I would get near Mexico, the border of Mexico, suddenly I've got like a hundred people. They got real nervous every time I tried to get out of the country or up to Canada, because I literally went from Mexico all the way up to Canada. Driving from San Diego. I left a note on my door that said I headed west from San Diego.

S: Headed west! That's a good idea.

DF: And then... crazy. Suddenly I had a new mailman and stuff, new neighbors. It was such a nightmare.

S: You still live at that same...

DF: No.

S: You live here now.

DF: Yeah.

S: You moved up here.

DF: Yeah, I'm like... I live with my friend.

G: I Western Unioned him some money.

DF: He saved my life, yeah. Robby saved my life. Without, like, yeah, I was at the point... They drove me to suicide.

S: No kidding!

DF: Yeah, I attempted it, and it didn't work, because I'm indestructible for some ungodly reason, I don't know why.

S: *[Laughs]*

DF: And he just goes, hey, you got to get out of there. He helped me with the money and stuff. My dad had sent me some money and I had my last unemployment check, and nothing in my house went stolen except that, and my license. And then my wallet disappeared and I'm, like, now what am I gonna do? Then he wired me the money. We would use the funniest... You have to ask a question, and there's an answer...

S: Oh yeah. To wire money?

DF: Yeah, to wire money. One was like, one of the best ones was, "What goes on the lamb?" "Mint jelly."

S: Oh yeah, yeah, yeah.

DF: That's a good one. But there was some funny things we used, like from Marx Brothers movies and stuff.

S: So this was three years ago that you moved up here?

DF: Yeah. I went from San Diego to Arizona, New Mexico, Texas, Louisiana, Mississippi, Tennessee, Missouri. Minnesota, where I have a twelve-year-old son. I went and dropped off everything I owned to my son so he could know his dad. And all my books and music and stuff. The relationship between me and her... she's remarried with kids and stuff. She's a really good mom.

S: That's your ex-wife?

DF: No, we had never even been married.

S: Oh, I see.

DF: When we broke up we had sex one more time and then she got pregnant. It was one of those situations. But she's a great mother and my son's a really... he makes movies. Twelve years old, he makes movies. Plays the violin. He's really incredible.

S: Where does he live?

DF: In Minnesota. But I went up there and dropped the stuff off to them and I didn't want to burden them with any of these problems, so I knew I needed to leave, plus we got into a horrible argument. Then from there I went through Nebraska, Wisconsin, down to Kansas. And my car broke down in Kansas. Then this guy offered me a job and I was going to buy a house for five grand. And I worked in a grain elevator and I tore down a train station, an old train station.

S: Wow.

DF: Yeah. But as soon as I got to Kansas, like Robby said, they left me alone. And it stopped. And since then I've been pretty much... Except recently, I was down in San Francisco during a big war protest, and they knew I was there.

S: Oh really.

DF: Yeah, which creeps me out. I think they possibly put something in my body.

S: How long ago was that?

DF: Fourth of July, around, yeah.

S: This was last year?

DF: Yeah, this was last year. But they followed me back up to my mom's house up in Eureka. She was an hour and a half out into the woods. It's literally, there's nothing out there. It couldn't be a more... It just so worked out that I've gone to really secure places and they've left me alone. I wasn't guilty of anything.

S: What's your full name?

DF: I'm in the book now.

S: Oh you are? What's your last name?

DF: Fuller, Dion. My dad, who... I've lived all over the country, I've lived a pretty eventful, adventurous life, but he was really concerned about this. My sister is in the Air Force, retired Air Force, she's a cop now, so I have a pretty, you know...

S: May I get your picture here? *[Pulls out a camera]*

DF: Oh no!

S: Okay. Now just turn your face to the left side. There, okay. Woops, wait a minute. I did the wrong thing. I'm sorry. Wait a minute. I pushed the wrong button.

G: Mention the message that you left before you...

DF: I don't remember.

G: You don't remember?

DF: No.

G: On the wall of the apartment?

DF: Oh...

S: Okay. Let's do it again. Now turn your face to the left just a little bit. Now the other way. Okay, just hold it there. I don't know why it's not...

DF: This is creeping me out.

S: It should be working.

DF: It's a derogatory comment. I didn't make life easy on myself at times, that's for sure.

S: Turn once more to the left. The reason is, I want to catch your ears. They're kind of interesting. It's kind of like one of my ears is crooked.

DF: Is it?

S: Yeah. *[Laughs]* Interesting.

DF: Yeah, that's... I'm glad to meet you. It just is incredible that you guys would have known each other anyways. I mean, this coincidence is amazing.

S: Well, your story really fascinates me. Are you writing something on this?

DF: Yeah, I'm trying to document—

S: I certainly want to get a copy—

DF: I'm trying to make... I'm trying to talk him into sitting down with me over like a four, five day period and then writing it all down. We're trying to figure out where to end it, because it just keeps going. And just the fact that you guys... you're a Freemason, he's a Freemason. That's incredible that you were both going to be... you both were going to be at the Lodge today. Anyway, it's just insane.

S: Yeah, right.

DF: Anyway, it's just insane, because I came down because you guys started correspondence over the email, and I came down specifically so we can meet. You guys were going to be in the same place anyway. That's insane. That's insane.

S: Yeah. Everything. The whole thing is pretty wild. And the fact that you had this experience...

DF: This experience. It's just, you know, too much.

G: I want to ask you, does it surprise you? Can you imagine that this occurred? I mean, before you said, well, I'm not sure if it's really even operational now. Does it surprise you that they would go to this extent and harass somebody in that way?

S: It doesn't surprise me because I know sometimes they will select a target who is vulnerable that they can use. They know they can't possibly get any leaks from them because if they do get leaks, people think they're crazy. They'll discredit them. They'll say, oh, he was arrested for this and that. All kinds of reasons to discredit them.

DF: Yeah, that's what's good about Robby being a college professor and everything, so it's not going to play out in the way in which they had hoped.

G: Certainly they looked at your background.

DF: Oh yeah. I had a book. They downloaded everything off my computer and stuff. They know what I do. But also, the stuff I write is insane. And, to be honest with you, I've been arrested and stuff, so what they think... But at the same time I'm more than willing to share with people what's happened to me. And I'll reveal my secrets before they can. So then that's good. They're not going to have as easy a time to discredit me as they think they're going to.

S: How we doing on the parking?

DF: That's a good question.

S: I would think we're about done. Are we about done here?

G: At some point I would like to meet with you, actually, just you and me, and we'll just talk about Freemasonry.

S: Yeah, sure. Maybe we better go retrieve our cars so we don't get tickets, huh?

DF: Yeah. Thank you.

S: I sure appreciate—

DF: No, thank you.

S: —having this discussion.

DF: Oh yeah.

S: I want to follow up with you later.

DF: Oh yeah. Any time we get to see a demonstration, I would love that.

[All four leave the office and begin walking across campus toward the parking lot]

S: My partner out in Hemet wants me to develop something real simple. He said do something real easy, something small, hand-held, that you can go around with. I said, sure, I can do that. I think I can do that. It won't cost me too much.

G: Are you having trouble with financing right now?

S: Well, to do the sophisticated stuff, yeah. We've done some lab experiments and things like that down in Hemet, and we've demonstrated the concept. It's not a question of the concept not working, it's more the degree of fidelity you can get to cut down on the—

DF: Yeah, a real sparkling mirror-type effect is what I saw.

S: Yeah. If you want the best displays you're going to have to put some extraordinary coatings on there, antireflective coatings on it. And that's the problem I run into. That starts to get expensive. I have a concept for a type of display that would prevent this, but to develop that display it would probably cost six thousand dollars or so. So you're writing an article or a book?

DF: A book.

S: An exposé?

DF: Well, yeah, in a way.

S: You might get a contract for a movie or something.

DF: I mean, gee, we'll be coming to you to do the special effects. *[Laughs]*

S: You remind me of one of the guys in the series *Taken*, where he was running all over the country, trying to get away from the government.

DF: That's what I did. It was ridiculous. There was a scene where I was in a parking lot in the middle of winter in Wisconsin and a woman was following me. And I cornered her with my van in the parking lot, and I was yelling at her, "Who do you work for?"

S: Was she good-looking?

DF: *[Laughs]* Nah. "*Why you following me?*" But it was so ridiculous. I'm like yelling at her in the parking lot. And anybody that was innocent would have called the cops on my behavior, but since they weren't—

S: We got a little rain since we went into the meeting, didn't we? Well, I hope your tapes and everything came out all right.

G: I'm still taping.

S: Oh, you're still taping?

G: Yeah.

DF: He's got a lot of writing on this stuff too.

G: I'll send you the Stephan Hoeller interview.

S: Oh, thank you. I'd love to see that.

G: In fact, I can just send you the link to it.

S: Okay. It's on the Internet?

G: Oh yeah, yeah. The whole thing is up there.

S: Good. Good. Did you download any material from my site?

G: Oh yeah, yeah, I printed it out.

S: A lot of stuff you should download. I haven't got my latest paper on there yet.

G: What's the latest paper?

S: The one that I presented back east at Charleston.

G: That would be very good.

S: But I can email that to you.

G: Yeah.

S: It's a large file. It's a PowerPoint file.

G: That's fine.

S: The reason I haven't put it on the site is because they asked me not to for a while, until they had made their presentation. But they made their presentation, and I could have gone ahead and put it on the site. I might do that now.

G: Did they say, "Can you wait 'til we stop torturing Dion Fuller?"

S: *[Laughs]* If I keep putting stuff on this site, I'm giving people ideas. I'm giving competitors ideas that they can take and run with, you know.

DF: Yeah. Is there any way you can get control of that? It's a real shame, because I know...

S: You say it's proprietary and you can even get it patented. I could get another extension on my patent or I could file for an amendment that would allow me to extend the patent.

DF: Everything that I've looked through on the Internet and stuff, the guy from Japan and the descriptions of all this stuff, based on my real-life experience with this stuff, when I read your thing, I'm like, this is it, this is what they're using. I know from a personal perspective, they stole your stuff. That's crap. And the way that the military works, it's like you said, they subcontract stuff out to companies and stuff and so...

S: Yeah.

DF: And there is the phenomenon of two separate people with a simultaneous idea, coming up with the same thing—but it does kind of sound like, when you had the meeting with those people, and they didn't get back to you, that just sounds shady.

S: Yeah, it is shady.

DF: They're not being forthcoming.

S: Sometimes they're presenting the image, well, we don't need you, you know.

G: Well, what do you think? Are you going to tell your partner about all this? What do you think he's going to say?

S: My partner in Hemet?

G: Yeah.

S: You mean about our interview?

G: Yeah, yeah, yeah.

S: Yeah, I'd like to, if you don't mind.

G: Oh yeah, sure, absolutely.

DF: It will eventually come out in a magazine.

S: You might like to interview my partner...

G: Absolutely.

S: Because he is a fascinating guy. He's from Russia.

G: Is he a Mason as well?

S: No, he's not a Mason, but he... he has an accent a little bit like...

G: Hoeller?

S: Dr. Hoeller, and he's got a good sense of humor and a real wit to him, and he used to work for the Russian navy, I believe. He is a civilian and he's also military. And he was telling me... He got interested and sympathetic toward my design, because he said, in World War II in Russia, that many times they would camouflage railway trains by painting on the roof. They would paint the railroad tracks on the roof of the train, so if you're looking down from above, all you see is the railroad tracks.

DF: That's funny.

S: You can't see the train moving. So he got real intrigued with it, and he's been my supporter, one of my sole supporters for quite a long time. And he's collaborated with me on a couple of papers. One we presented at the American Physical Society, at their Centennial. This was back in '99, I believe. And then he helped me on my most recent papers too. He sort of went over some of the calculations and things, confirmed what I said was true and reliable. He's very interesting. He's got a laboratory where he does testing for various companies, and the government. He even tests some items for the government.

DF: He lives in Hemet?

S: In Hemet, right. He has a little laboratory out there. An excellent place for us to do work if we got a contract. It would be a good isolated place to do work.

DF: Yeah, Hemet is pretty much no-man's land.

S: So I took the Navy out there, these Navy people. I escorted them out there and gave them a tour of that lab. They were quite impressed with it. They liked it. But we haven't got any contract yet.

G: And he lives out in Hemet?

S: He lives in Hemet, right. If you have a Saturday sometime when you're free, we could schedule a trip out there.

G: Oh yeah, that would be fantastic.

S: You could interview him, say, about his relationship to Project Chameleo and his experience with that—

G: Yeah, absolutely.

S: —technology and all of that.

G: That would make it even more complete.

S: Yeah. He'd give you a different… it would give you the perspective from a physicist's angle.

G: Yeah.

[All four arrive at the parking lot]

S: He's an eminent physicist, well recognized all over the world. Did we get any tickets yet?

DF: No.

S: Well, it's sure been nice meeting you, Dion and Robert and Melissa. And thank you for all your...

DF: Thank you for lunch, and thank you.

S: Oh, you're welcome.

DF: My honor, thank you.

S: Sure, we'll do this again before long.

G: Thanks a lot.

[Schowengerdt gets into his car and drives away.]

G: Well, it's four o'clock on the dot.

DF: That was pretty insane. What did you think of that?

MO: That was pretty awesome.

G: I told you he's an interesting guy.

DF: To say the least. Incredible stuff, huh? I mean... yeah, that's just crazy. It confirms everything I already knew, which is nice. He's a great guy too. He's totally down-to-earth. Because my story could scare the shit out of somebody.

MO: Did you get the feeling when you first started telling the story... I couldn't tell if he was going to get really pissed off or not. The first second, his face just sort of changed a little bit.

G: Really? Because I wasn't looking at him, I was looking at Dion.

MO: But then he was totally fine, but for a second he seemed like, what are you about to spring on me, like? Are you making fun of me or are you... you know? Something was weird, I think.

DF: No, no. Same thing, same thing...

G: Maybe he might have thought you were accusing him of something.

DF: No, you're right, because when we were at the restaurant I got the same response and I just went on to something else. Here I pursued it.

G: Yeah, but then I think he quickly realized you weren't accusing, you were just sharing information.

MO: Just for a second I wasn't sure what was about to happen. *[END]*

✳ ✳ ✳

By the time the interview was over I felt as if I had just done a workout or something. I was amazed that almost everything Dion had ever told me had been confirmed, in technical terms, by the very man who invented the technology the U.S. military had used to torture Dion in the first place. After Dion, Melissa, and I said goodbye to Richard at four o'clock, the three of us wandered over to the Indian burial ground located on the edge of campus. As we wandered around the burial grounds talking about what had just gone down, I began thinking about the power of synchronicity. The military—and/or paramilitary—team in San Diego responsible for this unique brand of psychological terrorism clearly needed experimental subjects to test their equipment on in real time. Their victims are people they deem to be A) useless bottom feeders and criminals and B) so marginalized and disconnected that no one will care about their experiences or believe them. Not even these Black Pajama Boys could predict that someone like Dion would have a friend who A) had already researched government mind control programs to the extent that I was personal friends with Walter Bowart, the man who first wrote about the subject in any depth at all, B) was a 32nd Degree Freemason, and C) was part of the same Lodge as the scientist from whom they had stolen the very technology they were using to experiment on

Dion and scores of other marginalized people existing on the fringes of society. Who could foresee that? Not even the most sophisticated supercomputer in the universe could predict a set of circumstances as bizarre as that. As the mythologist Joseph Campbell once said, "Synchronicity is the universe's way of showing you that you're on the right path." Indeed.

Oh, by the way, any doubts I had about the capacity of the U.S. government to ignore the Constitution and torture marginalized people with unrelenting ruthlessness was counteracted by the fact that we were standing in the middle of an Indian burial ground. By the time this reaches print, I wouldn't be surprised if there's a McDonald's sitting on top of it. (I wouldn't like to work under *that* particular pair of Golden Arches.)

Eventually Melissa had to leave. She seemed excited by the entire interview, and said she thought she'd gotten some good shots. She said she'd call me to look at them soon. Now it was just me and Dion. I could tell he was both relieved and frightened at the same time: relieved that he wasn't crazy (I think he himself hadn't been certain), and frightened that everything had happened exactly as he thought it had. I think he would've preferred to be crazy. It would be easier living in a world where Dion made up invisible midgets off the top of his head as opposed to one in which the U.S. government actually manufactures them out of hermetic, alchemical paste.

The two of us went to a local fast food joint, bought a couple of teriyaki chicken bowls, and continued to talk about the interview for a couple more hours. I thought Dion would be a little more relieved at having his experiences confirmed, but he seemed on edge. When we parted that evening, he even appeared to be a little depressed.

What happened from this point on would be another book in itself. In fact, it is. It's called *The Opposite of Foolproof* and though it's ostensibly a novel, it pretty much accurately depicts all the chaos that occurred during the next four weeks.

The relationship between me and Dion disintegrated as my relationship with Melissa grew deeper. Though Laurie would continue to pop back into my life without warning like a papier-mâché phantom in a carnival scare show, for the most part I had turned my back on her—including her memory. Melissa was what I had really wanted the entire time I had been trying to adjust to the brutal ups and downs with Laurie.

For some reason the interview with Richard seemed to drive Dion over the deep end. At one point he announced he was going to return to San Diego to get revenge on the bastards that had tried to kill him and to win Jessica's heart back, like a modern day Odysseus (if Odysseus lived in a windowless black van and was addicted to heroin). Just as soon as he announced this, however, he promptly laid down stakes and set up a semi-permanent drug clinic in the parking space right outside my apartment building. One minute he told me in all earnestness that he wasn't carrying any drugs on him at all, then a few days later proceeded to show me boxes and boxes of marijuana and packs of black tar heroin stored in the back of his van—without any acknowledgment of the contradiction. He claimed his movements and phone calls were still being monitored by the U.S. government, but he would go out of his way to use *my* phone to make drug deals. He started selling drugs to everyone who lived in my apartment building by going door to door like a salesman trying to pawn off subscriptions for the *L.A. Times*. At one point he asked me if I could—you know, just casually, maybe after class—ask some of my hipper students if they wanted to buy drugs at a cheap price from my good friend Dion. When I refused to do that, he stopped Melissa in the hall and asked her if she could help sell some drugs on the CSULB campus. He picked up two homeless teenagers and allowed them to live in the van with him. Within weeks they were both hooked on heroin, not long after which the little ragamuffins stole the van right out from under him along with all the drugs inside it. He kept insisting he was going to write a book about his strange experiences, and if he had done that instead of committing random acts of nonsense everything would have turned out fine—for him and everyone else associated with him. Instead he allowed two street urchins to steal the manuscript-in-progress along with all his photographs and the various items he'd borrowed from me during the preceding four weeks. Now he was truly homeless again, sitting cross-legged on my living room floor, bawling his head off over the injustice of the universe... as if he hadn't caused all this trouble in the first place! It's amazing how one drunken bicycle ride can change an entire life—not just *his* life, but mine as well. If he'd never broken his leg he never would've stopped working and if he'd never stopped working he wouldn't have opened his home to an AWOL arms smuggler wanted by Naval Intelligence and if he hadn't opened his apartment to an AWOL arms smuggler wanted by Naval Intelligence he wouldn't have... well, we could on and on, couldn't we? What's the point?

I'd had enough. I felt as if I'd gone above and beyond the call of duty. I'd dealt with more madness in the past four years (from 2003 to 2006) than most people had to deal with in a lifetime. I bought Dion a Greyhound bus ticket and put him on a bus for Humboldt, where his mother lived. Right before he left my apartment for the last time, out of his mind on any number of different drugs I would never even attempt to identify, he said, in kind of hangdog manner, "Well... this is probably the last time we'll ever see each other, man."

"Somehow I doubt that," I said.

The last I heard he was living in San Francisco and eking out some kind of meager existence as a painter. On December 24, 2008, he wrote me the following email:

Merry Fuckin' Christmas

I'm in San Francisco, Broke and trying to get back in Art school. I did a semester in Santa Cruz, Cabrillo junior College. Can you check my grades I never picked them up. Well how much money do I owe you? Tell Your Folks Happy Holidays, I'll try calling them.

Go FAsT-Stay SCARY
dIon

I didn't respond. I loved how he told me he was broke, then asked me how much money he owed me. By this point I'd forgotten he even owed me money. Perhaps he thought that's why we weren't communicating anymore? I don't care about lost money. I'm far more stingy about lost time.

20.

On March 25, 2010, I received a phone call from Dion (how he got my phone number I have no idea). He'd left a message on my machine. In the message he told me he had attempted to call my parents to wish them a Happy New Year a few months before, but was surprised to discover that their phone number had been disconnected. He just wanted to know that they were okay. I wrote him the following email:

Just received your message tonight. Mom & Dad both moved to Wisconsin back in the Spring of 2008. In terms of their health (both mental and otherwise), they've never been better. —RG

On March 28, he wrote back:

Hey Robby,

That's Great! I was blown away. A complete unwavering piece of reality suddenly disappeared. It was like looking up and watching the moon vanish. I hope mentally and otherwise that you are better then (than) ever as well.

Happy Birthday! You old fart. How is work? I've got some stuff to send you.

I live in SF. In the Cadillac Hotel, the first hotel built in the Tenderloin in 1907, Muhammad Ali, Cassius Clay trained in its Gym and Dead Head that's really dead Jerry Garcia once lived there. Now it's me and my dog Bruce who live next door to a transsexual that's a dead ringer for the Octo-Mom. I'm being paid to Paint the Janitor's wife's portrait. It is really not as easy as it sounds. She registers 7.7 on the Fugly Scale. That's fucking funny. I should write for TV.

Anyway, Tell your Mom and Pop, Hello and that I miss them and that I'm happy they escaped Torrance. Give me a call sometime (415) 409-0913
fuck Humanity!

Dion

p.s. "My wife glued her asshole shut"
I over heard that while walking the dog.

Though a couple of my friends (who've known Dion for years) suggested I not communicate with Dion any further than this, I decided to email him an autobiographical story I'd recently written entitled "Evicted" that explained in exact detail why my parents were forced to move to Wisconsin. I thought I would hear back from him about it, but never did. Perhaps he was evicted himself before he could check the email. Who knows?

✻ ✻ ✻

A condensed version of my interview with Richard Schowengerdt was eventually published in the March 2007 issue of
UFO Magazine.[24] I had submitted the article to a whole variety

---

24. The interview appeared under the title "To See the Invisible Man," and
was preceded by the following introduction...

"The title of this interview serves several purposes at once, both literal and
metaphorical. Though it's a reference to a classic 1963 science fiction story
written by Robert Silverberg, the subject matter of this interview is hardly
science fictional. Richard Schowengerdt, founder of Project Chameleo, has
successfully developed a technique to make men invisible. A team of U.S./
British researchers at Duke University in North Carolina and a Tokyo professor named Susumu Tachi have recently attracted a great deal of media
attention due to having made very similar claims. However, whereas the
invisibility techniques of both Tachi and the Duke University team remain
rather limited in scope, Schowengerdt's has the potential to revolutionize
covert warfare—if, in fact, such a contingency hasn't already occurred.

"Over the past five decades Schowengerdt has been quietly responsible for
a number of innovations in the field of electromagnetics while working
under the auspices of the U.S. military at installations such as the Navy
Metrology Engineering Center (MEC), the Naval Sea Systems Command
Technical Representative Office (NAVSEA TECHREP AEGIS), and with
the Naval Reserve at the Miramar Naval Air Station and the Naval Air
Station, Pt. Mugu, among several other installations. At MEC he was in-
strumental in developing the first digital voltmeters in concert with Industry, as well as directing nuclear magnetic and Josephson's Effect research
with the National Bureau of Standards, now designated as the National
Institute of Standards and Technology (NIST). Furthermore, when he
worked for the NAVSEA TECHREP he pioneered a concept for closed-
loop testing of guided missiles that reduced the need for excessive missile
firings on a test range. Currently, he is working at the El Segundo division
of Northrop Grumman Corporation (NGC), where he is responsible for
engineering surveillance of the NGC workshare on the EA-18G Growler,
a new fighter aircraft designed to replace the EA-6 Prowler, the primary
electronic warfare aircraft used by both the Navy and the Marines.

"Schowengerdt's private experiments with electro-optical camouflage
began in 1987, but it wasn't until 1993 that he launched Project Chameleo. He finally secured Patent No. 5,307,162 entitled 'Cloaking Using
Optoelectronically Controlled Camouflage' on April 26, 1994. Later he
teamed up with an associate in Hemet, CA, Dr. Lev Berger, and performed some tests and simulations involving Project Chameleo technology, culminating in the presentation of a paper, 'Physical Aspects of
Electro-Optical Camouflage,' at the American Physical Society Centennial in Atlanta on March 23, 1999. In February 2005 he presented
another paper titled 'Innovations in Electro-Optical Camouflage—

of magazines—*Wired*, for example—that I thought might be receptive to it, since they had published articles in the past about other ostensible breakthroughs in invisibility technology. Not one of these magazines ever responded, which is unusual. Never before has a magazine not responded to something I submitted to them, no matter how insane or off-the-wall it was. At the very least I should have received a form letter saying, "No, thank you." That's just standard operating procedure. But it was as if they had never received it. The only reason *UFO Magazine* did receive the manuscript was because I had the personal email address of the editor (thanks to a friend).

The interview was published along with Melissa's photos. I received almost no reaction to it whatsoever except for a strange phone call and email exchange in December of 2007:

> *Hello Robert,*
>
> > *My name is Linda Ciarimboli.*
> > *I am working for Paramount Pictures on the new future soldier movie, "Dark Sky." I have been asked to research making a uniform that blends in with it's [sic] surroundings. I found your article on project Chameleo.*
> >
> > *Does the fabric with lenses exist now or is it still being developed? Can you point me in the direction I need to go to get some?*
>
> > *Thank you for your time.*
> > *Linda Ciarimboli*

PROJECT CHAMELEO' at a Military Sensing Symposium at SPAWAR, Charleston, South Carolina. His pioneering experiments with optical camouflage continue to this day.

"And yet, despite these and other accomplishments, few people have ever heard of Schowengerdt's work. To correct this oversight, I decided to shine a spotlight on the Invisible Man himself. I recently sat down with Mr. Schowengerdt on the campus of CSU Long Beach and talked with him for three hours about the possible effects his technology will have not only on the military, but on everyday life."

She left a phone number. I called the number, but she wasn't in at the time. So I left my office number with her secretary. The next day I received the following email:

> *Good morning Robert,*
>
> *I got your email last night and phone memos this morning. I get in at 7am and work here until 6pmish. I don't want to interfere with your schedule so please call me on the office phone or on my cell after 6pm 310-871-_____. Office phone number here is 310-821-_____.*
>
> *Dark Sky is a Paramount pictures film about future soldiers. It will release in 2009 so I have to be sure to find the cutting edge or we will be behind the times when the film releases. Your article was passed to me by the designer who got it from the producer I believe. The powers that be want to use the invisible technology in a scene or two.*
>
> *I am a costume engineer for the film business. I develop new technologies and products for superhero and sci fi type applications. Batman, Superman, Spiderman, etc. I can give you more info when we speak.*
>
> *Thank you*
> *Linda C*

We finally connected on the phone, and she told me the film would involve invisible soldiers, so she wanted the technology to look realistic. I assumed that she wanted Richard to act as some kind of technical advisor to the film, so I passed along his personal phone number to her. I asked her how she had come across the article, and she said that the producer of the film often scours *UFO Magazine* for ideas. Most average people assume that publications like *UFO Magazine* are read by poor mental rejects wearing tinfoil hats, but the fact is that rich mental rejects in Hollywood are constantly ripping off these magazines because "conspiracy theories" can't be copyrighted. Case in point.

Linda C. called Richard, and instead of asking him to be a technical advisor on the film she asked him if she could buy a couple of invisibility suits from him. As if this weren't a theoretical proposition. As if he had a stack of these things sitting in his closet! If Richard had the money to manufacture them himself he wouldn't have had to pitch the project to the Navy for possible funding. Richard explained to her that his technology would work in theory, but he would need a vast budget to actually manufacture one of these suits. He said, however, that he could lend his expertise to the project in terms of letting them know how the suits should look and operate, etc. Linda C. seemed interested until Richard wanted to know if he would receive any money or credit for his contributions. She immediately said, "No," but after all it was a big Hollywood movie and wouldn't it be neat to be involved in something like that? Richard told her to take a hike. If Linda really wanted to see some invisibility technology in action, I would've told her to contact a certain Special Agent at NCIS and tell her she knew where her goggles were hiding.

I later discovered that the title *Dark Sky* was just a cover. The real title of the movie was *G.I. Joe: The Rise of Cobra*. The finished film did indeed involve invisibility suits, but that wasn't a major focus. I suspect the real audience for films like this are the impressionable young children who will see it, think it's real neat to grow up and be an invisible soldier, and then when the military wheels out their new invisibility technology within the next fifteen to twenty years, all those now-grown men will be eager to sign up and get their opaque asses blown off in some godforsaken desert at the back end of the world. "But, hey, at least I got to be invisible for a couple of seconds! Go, G.I. Team, go!" Jesus. If only those kids knew what those suits were originally used for: diminutive perverts jerking off while playing pranks on meth addicts in Pacific Beach. "Go, G.I. Team, go! The threat has been eliminated!" No, Joe, it hasn't. The threat's living in San Francisco, walking his dog Bruce, collecting a steady stream of crazy checks thanks to your weird shenanigans, and painting the janitor's 7.7 fugly wife for a living while some guy's wife is gluing her asshole shut.

Mission accomplished.

EPILOGUE

"At the risk of boring you, I'd like to ask a question.
What's goin' on here?"
—*The Navy vs. the Night Monsters*, 1966

The last time I saw Richard Schowengerdt (as I write this on July 30, 2010) was about six weeks ago on the evening of June 16. My friend and colleague, Randy Koppang, was scheduled to deliver a lecture in Costa Mesa, not far from Richard's house, about his recent book *Camouflage Through Limited Disclosure: Deconstructing a Cover-Up of the Extraterrestrial Presence* (Book Tree, 2007). I was eager to know what Richard thought about Randy's theories, as I believe Richard's research overlaps with the subject matter of Randy's book.

In Part Two of his book, Randy offers a brief but illuminating interview with Melinda Leslie, one of the best-kept secrets in the UFO field. Leslie claims to be an alien abductee who, following her various close encounters, has been repeatedly *reabducted* by military-style commandos and interrogated by them with cavalier interrogation techniques that sound suspiciously similar to those used by private corporations such as Blackwater, the company responsible for the illegal interrogation of suspected terrorists in Abu Ghraib and other concentration camp–style prisons in the Middle East. After reading Leslie's highly convincing testimony, which does not lack corroborating evidence, one wonders if Blackwater has been contracted to handle interrogations of a *different* kind on top of their rather lucrative deal in the Middle East.

Only a few hours before Randy's lecture began, I found myself sitting in a Mimi's Café in Costa Mesa eating dinner with Melinda. We were discussing Richard's research into invisibility technology. I told her briefly about Dion's experiences in San Diego, then told her that when Dion and I spoke to Richard for the first time we were surprised when he mentioned that representatives of a San Diego corporation had visited Richard and vacuumed up all his knowledge before returning to home base. I told her I thought this particular corporation might be the key to the whole mystery. Without even a second's hesitation, Melinda said, "Was it SAIC?"

"Yeah," I said. "SAIC headquarters isn't far from where Dion lived in San Diego, and they're the ones who interrogated Richard about his technology over ten years before all the crazy shit went down with Dion. How'd you know it was them?"

Melinda shrugged. "I've been researching the MILAB[25] phenomenon for almost two decades now," she said. "I've interviewed dozens and dozens and dozens of alien abductees who've been kidnapped and interrogated by government types. You wouldn't believe how many of these people are in some way connected to SAIC. It comes up over and over again. Not just once or twice, but *over and over again*. They're definitely tied into this MILAB phenomena somehow."

The reason I wanted to publish my interview with Richard in the pages of *UFO Magazine* was to give some of these "abductees," or MILAB survivors, the chance to read Richard's words and realize that what they interpreted as an extraterrestrial presence may, in fact, have derived from a far more human source. This doesn't mean *all* UFO phenomena can be swept under the rug and explained away in this relatively mundane fashion, but I suspect some celebrated alien abduction cases were, indeed, covers for illicit mind control experiments being

---

25. Military Abduction.

conducted by groups funded by U.S. intelligence agencies. Dion himself told me on several occasions that he saw beings in his San Diego apartment that, for all intents and purposes, looked and behaved in a decidedly inhuman fashion. If not for his skeptical nature and his association with me—as well as his subsequent meeting with Richard—who knows? He might have concluded that they *were* extraterrestrial in origin.

I'm not the first person to suggest such a possibility with regard to the alien abduction phenomenon. The first were Jacques Vallee in his 1979 book *Messengers of Deception: UFO Contacts and Cults* (a theory he later expanded on in his 1991 book *Revelations: Alien Contact and Human Deception*) and Martin Cannon in his 1996 book *The Controllers: A New Hypothesis of Alien Abduction*. Journalist Walter Bowart suggested the same possibility in the 1994 revised edition of his groundbreaking book *Operation Mind Control*.

In fact, Walter Bowart was the one who first mentioned SAIC to me back in 2003 and suggested there could be a connection between them and recent breakthroughs in mind control technology. That Melinda would mention the same corporation now, with only the slightest prompting, suggested to me that Bowart was correct. Indeed, Bowart was correct about a lot of the controversial research he first included in his book *Operation Mind Control* way back in 1978, but how correct he was probably won't become apparent to the mainstream media until many years from now. Perhaps after it's too late.

SAIC (Science Applications International Corporation) was founded during the Summer of Love in 1969 by J. Robert Beyster and a small group of mad scientists. According to their official website:

> SAIC's national security efforts reach across all branches of the military and support the full spectrum of military operations—from peace keeping and humanitarian mis-

sions to major conflicts. SAIC also helps the U.S. Department of Defense, the FBI, and other agencies combat terrorism, cybercrime, and the proliferation of weapons of mass destruction.

Whether helping the U.S. Army develop systems to enhance night vision capabilities, helping the Air Force train far-flung units worldwide via the Internet, or helping the Space and Naval Warfare Systems Command achieve seamless, secure multimedia connectivity for Navy ship and shore communications, SAIC's technical expertise enables U.S. forces to be safer and smarter.

SAIC also develops products and applied technologies which aid in anti-terrorism and Homeland Security efforts. Our anti-terrorism products range from vehicle and cargo inspection systems for contraband detection, portable X-ray inspection systems to inspect suspect packages for explosives, to software that assesses the consequences of technological and natural disasters to population, resources and infrastructure.

Somehow they neglected to mention that whole invisible midget thing or targeting innocent American civilians as guinea pigs, but that's okay. When it comes to National Security, some nefarious deeds are just necessary, no matter how ugly they might seem. Hey, you can't make an omelet without breaking a few eggs, right? SAIC is currently engaged in the utopian endeavor of creating the Perfect Omelet. And some eggs just don't fit into the recipe. Such eggs are known as "statistical outliers," better off swept to the side of the kitchen floor rather than allowed to contaminate the Omelet.

Pundits on both the Left and the Right decry the encroaching fascism they see on the horizon. Liberals accuse conservatives of being Nazis and conservatives toss the same accusations at liberals. Earlier today, on 7-30-10, I heard Rush

Limbaugh, of all people, urging everyone in his radio audience to go rent *The Lives of Others*, a film about the fascist surveillance culture of East Germany a few years before the fall of the Berlin Wall, in order to warn everyone of the nightmare he sees descending upon him and his fellow conservatives in the United States.

Most people, whether on the Left or the Right or in the middle of the political spectrum, just want to be left alone, to be free to pursue their own interests without needless obstructions. We all see the nightmare being formed around us, we all hate it, and yet it just keeps on coming. If everyone were to band together and recognize the simple fact that there are groups of people in this country (and perhaps all over the world) with far too much money on their hands and the perverse need to stalk and torture anyone who violates their myopic view of how human beings should think and behave, perhaps we could all do something to stop this dystopia from ruining our own lives and the lives of people we know... the lives of others.

Because the lives of others will soon become *our own* lives if we don't stop the nightmare before it grows out of anyone's control. And soon we will be living in Surveillance City, and no one will walk the streets unwatched. And no one will care anyway, because we will all have lost the ability to care about anything except working for the benefit of The Machine.

Note: the corporate headquarters of SAIC is located only eight miles from Dion's Pacific Beach apartment, a mere thirteen minute drive.[26]

SAIC also maintains an office on Camp Pendleton.

---

26. Update: SAIC's headquarters is now in McLean, VA. On 9-27-13 SAIC changed its name to LEIDOS, then spun off its IT Services Division into a new company that retained the name SAIC.

✳ ✳ ✳

In 2006, after the initial interview, Richard invited me and Melissa (the aforementioned photographer who soon became my girlfriend) to interview his partner, the celebrated physicist Dr. Lev Burger, in the professor's laboratory in Hemet. Melissa and I were taken on the same guided tour that the representatives from the Navy must have received back in 2005. Afterward, Richard invited us to have dinner with him and his wife, Emiko, at his lovely home in Costa Mesa. At one point during the drive back from Hemet, Emiko claimed she was psychic, but not in a boastful way; she just said it casually, as you might if you were telling someone you were an auto mechanic. It was just a fact of life for her.

Later that night, as we were leaving their house, Emiko took Melissa's hand and said, in a very matter-of-fact way, "We'll see you again after you're married and have the baby." We both laughed nervously. Melissa and I had no such plans at that point.

A few months later, several weeks after my thirty-fifth birthday, Melissa gave me a gift: a silver Masonic pocket watch with a note inside that read: I'M LATE. At first I thought she meant she was late with the gift, since my birthday had already come and gone.

"No, no," she said, "I'm *late*."

That's when it struck me.

When our daughter was born in January of 2007, Melissa and I decided to name her Olivia Emiko Guffey, giving Richard's wife proper credit for predicting a course of events that not even Melissa and I had seen coming.

✳ ✳ ✳

I had always intended to write about these experiences in detail, even if only in the form of a brief essay, but events out of my control forced me to push such plans aside. I had first tried to write it all down back when it was first happening. I started an article called "The Curious Affair of the Night Vision Goggles." I wrote about five hundred words, then stopped.

Very rarely do I *not* finish something I've begun writing. Even if I think it's horrible, I generally have to wrap it up. But with this piece I just pushed it aside and didn't think about it again. It was all too much. Too disturbing. I didn't want to dwell on it anymore. I lost myself in other endeavors. That brief, unfinished fragment was long forgotten until the spring of 2010 when I was teaching a class called The Literature of Science Fiction at CSU Long Beach.

I had just finished a lecture, a ridiculously *long* one in which I traced the history of the science fiction genre from 2000 B.C. to 1939 A.D., when a student named Francesco stopped me outside the classroom and asked me a question about invisibility. Francesco often asked me strange questions after class, but this one particularly caught my interest. He said, "If something becomes true, can it really be considered science fiction anymore? Like invisibility. Is that science fiction? I heard about some mad inventor in Japan who's whipped up some kind of invisibility cloak. Was that just bullshit? Have you heard about that?"

"I think you're talking about Professor Susumu Tachi," I said. "But I can do you one better than *that*." And I proceeded to tell him an abbreviated version of the story you've just finished reading. A couple of days later, during a lull in my typically peripatetic lecture, Francesco asked me to repeat the story to the entire class. I did. Some of them were skeptical, others not so much.

One student named Jason, who was in the Army and would often come to class wearing his full uniform, raised his hand and said, "Sir" (he was always very polite), "I think everything you said is very true because during our training, our superior gave us a whole rundown on S-4 and Area 51 and the experiments they're conducting out there. Of course, I can't tell you *exactly* what he said, as we were told not to repeat it, but I can say that everything you've just said ties into what I heard." Some of the other students seemed a little less skeptical after that.

The next day, I ran into Jason in the hallway. Jason said to me, "Sir, I'd just like to tell you how much I'm enjoying your Science Fiction class. It's refreshing to be experiencing military training while enrolled in your class as well. You see, it seems to me that your class—in fact, science fiction in general—is all about asking questions, whereas the military is all about *not* asking questions."

As mentioned before, the legendary science fiction writer Theodore Sturgeon once gave the following advice to all budding science fiction writers: "Ask the next question."

That's an excellent piece of advice for anyone, not just novelists with wild imaginations.

It still fascinates me that, if not for Francesco's random question, I may never have put aside my other writing projects to finally document this mad tale. "Synchronicity is the universe's way of showing you that you're on the right path."

Synchronicity. Happenstance. Fate. Destiny. Ineffable concepts unalterable by The Machine or the little men trying desperately to keep it running.

Perhaps it wasn't the right time to document this story back when I first attempted it in 2003 when I was still in the eye of the storm. Perhaps *now* is the right time.

Time to start asking the next question.

✳ ✳ ✳

UPDATE

Almost two years have passed since I wrote the previous sentence. In the interim, Dion has gotten in touch with me on several occasions. I've had lengthy conversations with him on the phone. He revealed to me that the harassment had, by and large, dissipated since his move to San Francisco. Significantly, the familiar pattern of being followed and watched would only recur when a protest erupted in the area—an anti-war rally, for example, or an Occupy Wall Street event. This being San Francisco, such events were not infrequent. On these days, Dion learned (the hard way) to stay indoors and keep well out of sight until all the excitement settled down.

In 2011, due to various reasons (mainly economic), Dion moved to a house located in the redwoods in Northern California. His mother lives in the middle of the forest next to a remote area appropriately named "the Lost Coast." Within weeks of his moving into the woods the harassment began again—with renewed intensity.

The High Strangeness kicked off when Dion began seeing classic, 1950s-style UFOs hovering over his mother's property, usually very late at night. He'd never before seen anything like these lights. They disobeyed all known laws of physics. After several weeks of documenting these sightings with his camera, he spotted a military-style drone flying low overhead in the night skies, well below the cloud cover. He watched in amazement as the drone seemed to release several lights that looked precisely like the UFOs he'd seen so often before.

It's important to note that this area of Humboldt County is filled with marijuana farmers who are protective of their crops, suspicious of outsiders, and trigger-happy to boot. As these drones began appearing in waves over the Mattole Valley, the local rednecks whipped out their rifles and began firing at the ominous craft. A rumor spread through the area that the farms would soon be raided by the FBI, or the ATF, or the DEA, or some similar government agency. One day a local resident spotted two "agents" (if, indeed, that's what they were) camped out on a nearby hillside; the "agents" appeared to be spying on the farmland below with a pair of binoculars.

The sunlight glinting off the binoculars gave them away. *En masse*, the local farmers ran up into the hills with their firearms, intent on blowing the intruders into jagged, bloody shards. The "agents," or whoever they were, beat their feet and scurried away like panicked vermin.

Later that night, the drone sightings in the valley grew even more intense. They were now so frequent that even Dion's mother (who had always been skeptical of Dion's San Diego experiences) admitted she couldn't explain their presence in the skies above her modest little home.

As of today, these weird sightings are still occurring. The old pattern has started up again. Dion often wakes in the middle of the night feeling nauseous, a metallic taste in his mouth, hearing the sounds of invisible intruders lurking around outside—sometimes even *inside*—his trailer located just behind his mother's house. His loyal dog Bruce often hears the sounds as well and skitters away in terror.

Of course, I allow for the possibility that the relentless harassment Dion suffered at the hands of the U.S. military has pushed him a few degrees (or more) past the already hazy border of Saneville. If he didn't start out crazy, he might very well be classified as such now.

Perhaps the appearance of these drones in the area is merely coincidental. After all, the U.S. government would naturally have an interest in keeping that part of the country under close scrutiny due to the amount of illegal substances grown there on a daily basis. But the reaction of the locals suggest otherwise. According to them, such bizarre activity has never occurred in the valley before. That is, not until Dion arrived.

It could be that whenever a "flagged" individual enters a zone of possible interest (e.g., an anti-war rally or the marijuana capital of the world), the robot spies and the invisible dwarves are wound up like toy soldiers and frogmarched in for the purpose of monitoring said individual up close and personal, just to see what the guy's up to, make sure he's not getting into any trouble.

But if the surveillance techniques of the U.S. government are this omnipotent, how does *any* terrorist ever get away with committing a crime, no matter how large or small? How is it possible that thousands upon thousands of tax payer dollars are being spent monitoring a harmless, forty-year-old freak who's living in his mother's backyard while *real* terrorists and miscreants are allowed to roam freely in and out of the United States at will?

Answer: who knows? In the final analysis, we might conclude that Dion annoyed Someone He Shouldn't Have, and that certain Someone placed him on a list no one wants to be on, and there Mr. Fuller will remain… forever.

I'm not sure how to help Dion or people like him, or if they even can be helped at this point. All I know is that once the remote-controlled Boot of Oppression comes hurtling out of the sky and onto your face, you're screwed.

Forever.

Perhaps the best we can hope for in this bright and shiny future called the twenty-first century is to hunker down in perpetual terror, never look each other in the eye, pray we don't accidentally piss off the wrong *Kommandant* with the authority to wipe out our lives and loved ones, and dwell in quaking fear of the gnomes and hobgoblins and invisible imps that still haunt the edges of our post-medieval minds. Though they now wear different costumes and vestments, the devils of old still plague our dreams and waking lives. And as our technology grows stronger and more intrusive, so too do the devils that govern the widgets and gears and microchips that made this brave new world possible in the first place.

Let's not be obtuse: we're dealing with a rule-crazy, Puritanical, hypocritical, Old Testament–style perception of reality that desperately needs to wipe out anything or anyone that is *Other. Different. Contrary.*

If devils don't exist today, rest assured these insane fuckers will figure out a way to bioengineer them into existence tomorrow. And when an unacceptable, transgressive thought invades the holy sanctity of your mind's eye, the devils will know.

The devils have always known.

Yes, the old daemons are being summoned from the mists of myth at the courtesy of an unlimited black budget and a military mindset that never forgets—nor forgives.

The tenets of the Old Testament, along with even more ancient grimoires, are now being deployed as strategic military weapons. What was mere religion yesterday will become covert policy tomorrow.

Fear God, demands our *novus ordo seclorum*. Fear the devils He created in his own image. Fear yourself. Fear the invisible.

That's the new National Anthem of the Disassociated States of Amurrrica. O say can you see, by the dawn's early light, the swarm of robot-eyes descending upon your hometown?

No, I doubt you can see them.

Don't bother watching the skies.

God's invisible, and so are the demon hordes controlled by His mighty hand. Shh. Wait. Remain still. Pray He's overlooked you. His titanic, arthritic finger is about to press down on a button and randomly destroy someone's life by placing them on THE LIST OF THE DAMNED. Forever. For no sane reason.

Perhaps it'll be my life. Perhaps it'll be yours.

Learn to enjoy the Fall.

Or think quick and figure out how to bend that damn finger back on itself until it *snaps*.

"There will be no more invisible people in America!"
—SENATOR HILLARY CLINTON, 1-8-08

Author Photograph © Melissa Guffey

ABOUT THE AUTHOR

A graduate of the famed Clarion Writers Workshop in Seattle, Robert Guffey is the author of a collection of novellas entitled *Spies & Saucers*. His first book of nonfiction, *Cryptoscatology: Conspiracy Theory as Art Form*, was published in 2012. He's written stories and articles for numerous magazines and anthologies, among them *Fortean Times*, *New Dawn*, and *The New York Review of Science Fiction*.

TECHNOCREEP: THE SURRENDER OF PRIVACY AND THE
CAPITALIZATION OF INTIMACY
Thomas P. Keenan
ISBN 978-1-939293-40-4 (paperback)
ISBN 978-1-939293-41-1 (e-book)

"*In* Technocreep, *Dr. Keenan explores some of the most troublesome privacy-invasive scenarios encountered on the web and offers users a number of excellent, practical ideas on how best to protect their privacy and identity online.*" —DR. ANN CAVOUKIAN,
INFORMATION AND PRIVACY COMMISSIONER OF ONTARIO

* * *

IVYLAND: A NOVEL
Miles Klee
ISBN 978-1-935928-61-4 (paperback)
ISBN 978-1-935928-62-1 (e-book)

"*A harsh, spastic novel about drug-addled misfits clawing their way through a wrecked future that feels disconcertingly familiar. As if that wasn't enough, it's also got evil caterpillars, flung jellyfish, great prose, new drugs, sharp jokes, a stolen ice cream truck and a miracle tree.*" —JUSTIN TAYLOR

O/R

FOR MORE INFORMATION, VISIT OUR
WEBSITE AT WWW.ORBOOKS.COM